华章IT | Information Technology

智能系统与技术丛书

Natural Language Processing with Java

Java自然语言处理

[美] 理查德 M. 里斯（Richard M. Reese） 著
邹伟 孙逢举 译

图书在版编目（CIP）数据

Java自然语言处理／（美）理查德 M. 里斯（Richard M. Reese）著；邹伟，孙逢举译．
—北京：机械工业出版社，2018.3（2018.11重印）
（智能系统与技术丛书）
书名原文：Natural Language Processing with Java

ISBN 978-7-111-59211-2

I. J… II. ①理… ②邹… ③孙… III. ① JAVA 语言 - 程序设计 ②自然语言处理
IV. ① TP312.8 ② TP391

中国版本图书馆 CIP 数据核字（2018）第 034180 号

本书版权登记号：图字 01-2015-7905

Richard M. Reese：*Natural Language Processing with Java*（ISBN：978-1-78439-179-9）．
Copyright © 2015 Packt Publishing. First published in the English language under the title "Natural Language Processing with Java".

All rights reserved.
Chinese simplified language edition published by China Machine Press.
Copyright © 2018 by China Machine Press.

本书中文简体字版由 Packt Publishing 授权机械工业出版社独家出版。未经出版者书面许可，不得以任何方式复制或抄袭本书内容。

Java 自然语言处理

出版发行：机械工业出版社（北京市西城区百万庄大街22号 邮政编码：100037）				
责任编辑：迟振春			责任校对：李秋荣	
印　　刷：中国电影出版社印刷厂			版　　次：2018年11月第1版第2次印刷	
开　　本：186mm×240mm　1/16			印　　张：13	
书　　号：ISBN 978-7-111-59211-2			定　　价：59.00元	

凡购本书，如有缺页、倒页、脱页，由本社发行部调换
客服热线：（010）88379426　88361066　　投稿热线：（010）88379604
购书热线：（010）68326294　88379649　68995259　　读者信箱：hzit@hzbook.com

版权所有 • 侵权必究
封底无防伪标均为盗版
本书法律顾问：北京大成律师事务所　韩光/邹晓东

译 者 序

THE TRANSLATOR'S WORDS

近些年来，人工智能领域的研究与应用已经成为全球性的科技热点，作为计算机科学领域与人工智能领域中的一个重要方向，而自然语言处理涉及语言学、计算机科学、数学等多种学科，必然也会受到越来越多的关注。因此，本书的出版正为其时，正当其用。

美国塔尔顿州立大学教授 Richard M Reese 的这本书是一部自然语言处理领域的著作。全书内容丰富，对自然语言处理的基础知识进行了全面描述与总结。本书详细介绍了自然语言处理的多种技术，包括 NLP 工具、文本分词、文本断句、词性判断、人物识别、文本分类、关系提取以及组合应用等，结合多个示例进行深入分析，并采用 Java 编程语言进行处理与结果分析。

翻译一本教科书式的英文原著，真的是一次非常难忘的经历，感谢机械工业出版社给我提供了这样的机会，我也尽量保持了原著的风格和行文特点。承蒙多位学者鼓励，译者有幸将国外优秀的原著介绍给广大读者，我在此表示衷心的感谢，同时感谢出版社编辑对文稿做了大量处理工作。无论您是工程师、科研工作者、学生，还是教师，希望在阅读本书后都能够有一定的收获。

虽然我已经竭尽全力还原原著的本意，也进行了多轮校稿和通读，但由于时间以及水平有限，难免有不当之处，敬请广大读者见谅并赐正。

邹 伟

2017 年 11 月 20 日

ABOUT THE AUTHOR
作者简介

Richard M. Reese 曾就职于学术界和工业界。他曾在电信和航天工业领域工作 17 年，期间曾担任研发、软件开发、监督和培训等多个职位。他目前任教于塔尔顿州立大学，运用他多年来积累的行业经验来完善他的课程。

Richard 曾出版过关于 Java 和 C 的书籍，他使用简洁易用的方法讨论主题，这些书籍包括《EJB 3.1 Cookbook》，有关 Java 7 和 Java 8 的新功能、Java 认证以及 jMonkey 引擎，以及一本关于 C 指针的书。

我要感谢我的女儿詹妮弗，因她发表了很多评论，并做出很大贡献。她的付出是无价的。

审校者简介

Suryaprakash C. V. 自 2009 年开始工作于 NLP 领域，他先后毕业于物理学及计算机科学专业。后来，他有机会进入他喜欢的领域（自然语言处理）工作。

目前，他在 Senseforth Technologies 公司担任技术顾问。

我要感谢同事们支持我的事业和工作。这在审稿过程中给了我很大帮助。

Evan Dempsey 是一名爱尔兰 Waterford 的软件开发人员，当他不再因兴趣与利益从事黑客工作后，他爱上了精酿啤酒、Common Lisp，并继续从事机器学习的前沿研究。他是许多开源项目的贡献者。

Anil Omanwar 是一个充满活力的人，他对于最热门的技术趋势和研究充满激情。他拥有超过 8 年的认知计算研究经验，先后从事过自然语言处理、机器学习、信息可视化等领域，下一个主要研究领域是文本分析。

他精通各种领域下的情感分析、问卷反馈、文本聚类、短语提取等技术，这些领域包括生命科学、制造业、零售业、电子商务、酒店业、客户关系、银行业和社交媒体等。

他目前与 IBM 实验室合作，主要项目为生命科学领域的 NLP 与 IBM Watson。他的研究目标是能够自动化关键手动步骤，并协助领域专家优化人机功能。

在业余时间，他喜欢公益、徒步、摄影和旅行。他随时准备迎接技术挑战。

Amitabh Sharma 是一名职业软件工程师。他曾在电信和商业分析领域的企业应用方面做过大量工作。他的工作专注于面向服务的架构、数据仓库和语言等，如 Java、Python 等。

PREFACE

前　　言

自然语言处理（NLP）已用于解决各种各样的问题，包括对搜索引擎的支持，对网页文本的总结与分类，以及结合机器学习技术解决诸如语音识别、查询分析等问题。它已经在任何包含有用信息的文件中使用。

NLP用于增强应用程序的实用性和功能，主要通过简化用户输入以及将文本转换成更加可用的形式来实现。实际上，NLP能够处理各种来源的文本，使用一系列核心NLP任务从文本中转化或提取信息。

本书重点介绍NLP应用中可能遇到的核心NLP任务，每个NLP任务都从问题的描述以及可应用领域开始。介绍每项任务中比较困难的问题，以便你能更好地理解问题。随后通过使用大量的Java技术和API来支持NLP任务。

本书涵盖内容

第1章解释了NLP的重要性和用法。本章以简单的例子来解释如何使用NLP技术。

第2章主要讨论标记化，标记化是使用更为先进的NLP技术的第一步，本章介绍了核心Java和Java NLP标记化API。

第3章证明句子边界消歧技术是一个重要的NLP任务。这一步是其他许多下游NLP任务的预处理步骤，其中文本元素不应跨越句子边界进行分隔。这样就可以确保所有短语都在一个句子中，并支持词性分析。

第4章涵盖了通常所说的命名实体识别。这个任务主要涉及识别人、地点和文本中相似的实体。该技术是处理查询和搜索的初始步骤。

第 5 章说明如何检测词性，词性是文本中的语法元素，例如名词和动词。识别这些元素是确定文本含义和检测文本内关系的重要步骤。

第 6 章证明文本分类对于垃圾邮件检测和情感分析等任务非常有用。此外，本章也对支持文本分类的 NLP 技术进行了调查和说明。

第 7 章演示解析树。解析树可应用于很多目的，其中包括信息提取。信息提取拥有这些元素之间关系的信息。通过一个实现简单查询的例子来说明这个过程。

第 8 章包含从各种类型的文件（如 PDF 和 Word 文件）中提取数据的技术。接下来主要介绍了如何将以前的 NLP 技术结合至一个管道中以解决更大的问题。

阅读本书的技术准备

Java SDK 7 用于说明 NLP 技术。各种 NLP API 是必需的并可以随时下载。IDE 可选择，并不做强制要求。

本书读者对象

对 NLP 技术感兴趣的、有 Java 经验的开发人员会发现这本书很有用。不需要事先具备 NLP 知识。

目　录

译者序
作者简介
审校者简介
前言

第 1 章　NLP 简介 ·········· 1
1.1　什么是 NLP ············ 2
1.2　为何使用 NLP ·········· 3
1.3　NLP 的难点 ············ 4
1.4　NLP 工具汇总 ·········· 5
　　1.4.1　Apache OpenNLP ········ 6
　　1.4.2　Stanford NLP ············ 7
　　1.4.3　LingPipe ················· 9
　　1.4.4　GATE ·················· 10
　　1.4.5　UIMA ·················· 10
1.5　文本处理概览 ·········· 10
　　1.5.1　文本分词 ··············· 11
　　1.5.2　文本断句 ··············· 12
　　1.5.3　人物识别 ··············· 14
　　1.5.4　词性判断 ··············· 16

　　1.5.5　文本分类 ··············· 17
　　1.5.6　关系提取 ··············· 18
　　1.5.7　方法组合 ··············· 20
1.6　理解 NLP 模型 ········· 20
　　1.6.1　明确目标 ··············· 20
　　1.6.2　选择模型 ··············· 21
　　1.6.3　构建、训练模型 ······· 21
　　1.6.4　验证模型 ··············· 22
　　1.6.5　使用模型 ··············· 22
1.7　准备数据 ·············· 22
1.8　本章小结 ·············· 24

第 2 章　文本分词 ·········· 25
2.1　理解文本分词 ·········· 25
2.2　什么是分词 ············ 26
2.3　一些简单的 Java 分词器 · 28
　　2.3.1　使用 Scanner 类 ········ 29
　　2.3.2　使用 split 方法 ········· 30
　　2.3.3　使用 BreakIterator 类 ···· 31
　　2.3.4　使用 StreamTokenizer 类 · 32
　　2.3.5　使用 StringTokenizer 类 · 34

2.3.6 使用Java核心分词法的性能考虑 ……… 34
2.4 NLP分词器的API ……… 34
　2.4.1 使用OpenNLPTokenizer类分词器 ……… 35
　2.4.2 使用Stanford分词器 ……… 37
　2.4.3 训练分词器进行文本分词 ……… 41
　2.4.4 分词器的比较 ……… 44
2.5 理解标准化处理 ……… 45
　2.5.1 转换为小写字母 ……… 45
　2.5.2 去除停用词 ……… 46
　2.5.3 词干化 ……… 49
　2.5.4 词形还原 ……… 51
　2.5.5 使用流水线进行标准化处理 ……… 54
2.6 本章小结 ……… 55

第3章 文本断句

3.1 SBD方法 ……… 56
3.2 SBD难在何处 ……… 57
3.3 理解LingPipe的HeuristicSentenceModel类的SBD规则 ……… 59
3.4 简单的Java SBD ……… 60
　3.4.1 使用正则表达式 ……… 60
　3.4.2 使用BreakIterator类 ……… 62
3.5 使用NLP API ……… 63
　3.5.1 使用OpenNLP ……… 64
　3.5.2 使用Stanford API ……… 66
　3.5.3 使用LingPipe ……… 74

3.6 训练文本断句模型 ……… 78
　3.6.1 使用训练好的模型 ……… 80
　3.6.2 使用SentenceDetector-Evaluator类评估模型 ……… 81
3.7 本章小结 ……… 82

第4章 人物识别

4.1 NER难在何处 ……… 84
4.2 NER的方法 ……… 84
　4.2.1 列表和正则表达式 ……… 85
　4.2.2 统计分类器 ……… 85
4.3 使用正则表达式进行NER ……… 86
　4.3.1 使用Java的正则表达式来寻找实体 ……… 86
　4.3.2 使用LingPipe的RegEx-Chunker类 ……… 88
4.4 使用NLP API ……… 89
　4.4.1 使用OpenNLP进行NER ……… 89
　4.4.2 使用Stanford API进行NER ……… 95
　4.4.3 使用LingPipe进行NER ……… 96
4.5 训练模型 ……… 100
4.6 本章小结 ……… 103

第5章 词性判断

5.1 词性标注 ……… 104
　5.1.1 词性标注器的重要性 ……… 107
　5.1.2 词性标注难在何处 ……… 107
5.2 使用NLP API ……… 109

5.2.1 使用 OpenNLP 词性标注器 ·········· 110
5.2.2 使用 Stanford 词性标注器 ·········· 118
5.2.3 使用 LingPipe 词性标注器 ·········· 125
5.2.4 训练 OpenNLP 词性标注模型 ·········· 129
5.3 本章小结 ·········· 131

第 6 章 文本分类 ·········· 132
6.1 文本分类问题 ·········· 132
6.2 情感分析介绍 ·········· 134
6.3 文本分类技术 ·········· 135
6.4 使用 API 进行文本分类 ·········· 136
    6.4.1 OpenNLP 的使用 ·········· 136
    6.4.2 Stanford API 的使用 ·········· 140
    6.4.3 使用 LingPipe 进行文本分类 ·········· 145
6.5 本章小结 ·········· 152

第 7 章 关系提取 ·········· 153
7.1 关系类型 ·········· 154
7.2 理解解析树 ·········· 155
7.3 关系提取的应用 ·········· 156
7.4 关系提取 ·········· 159

7.5 使用 NLP API ·········· 159
    7.5.1 OpenNLP 的使用 ·········· 159
    7.5.2 使用 Stanford API ·········· 162
    7.5.3 判断共指消解的实体 ·········· 166
7.6 问答系统的关系提取 ·········· 168
    7.6.1 判断单词依赖关系 ·········· 169
    7.6.2 判断问题类型 ·········· 170
    7.6.3 搜索答案 ·········· 171
7.7 本章小结 ·········· 173

第 8 章 方法组合 ·········· 174
8.1 准备数据 ·········· 175
    8.1.1 使用 Boilerpipe 从 HTML 中提取文本 ·········· 175
    8.1.2 使用 POI 从 Word 文档中提取文本 ·········· 177
    8.1.3 使用 PDFBox 从 PDF 文档中提取文本 ·········· 181
8.2 流水线 ·········· 182
    8.2.1 使用 Stanford 流水线 ·········· 182
    8.2.2 在 Standford 流水线中使用多核处理器 ·········· 187
8.3 创建一个文本搜索的流水线 ·········· 188
8.4 本章小结 ·········· 193

CHAPTER 1

第 1 章

NLP 简介

自然语言处理（NLP）是一个宽泛的主题，它以借助计算机分析自然语言为核心，主要涉及语音处理、关系结构提取、文档分类、文本摘要等任务。不过，这些看似各异的任务都依赖于一些基本技术，包括分词、断句、分类和关系提取等，而本书也更侧重于这些基本技术的研究。首先，本章将详细讨论什么是 NLP，为何 NLP 非常重要，以及 NLP 的具体应用领域有哪些。

很多语言和工具都支持 NLP 任务。本书主要讨论 Java 语言以及各种 Java API 如何支持 NLP。本章首先介绍一些常用的 API，包括 Apache 的 OpenNLP、斯坦福的 NLP 库，以及 LingPipe 和 GATE 等。

接下来进一步分析前面提到的那些 NLP 基本技术。本书将基于 NLP API 介绍这些技术的基本原理及其具体使用方法。很多技术都会使用一些模型，这些模型可以看作一组规则，这些规则用于执行分词等任务。它们通常由从文件实例化的类表示。最后会说明如何为支持 NLP 任务准备数据。

NLP 并不简单。虽然有些问题可以相对简单地解决，但大多数问题都需要使用非常复杂的技术。本书仅使读者对 NLP 处理技术有初步认识，使其在处理具体问题时能够使用相应的技术。

NLP 是一个非常复杂的领域，本书通过 Java 实现一些核心的 NLP 任务，以帮助读者略窥自然语言处理中冰山的一角。在书中，通过 Java SE SDK 和 OpenNLP、Stanford

NLP等开源库展示了NLP的一些基本技术。使用这些库以前，需要将一些API JAR文件关联到相关API的项目中。关于这些库的说明可以参照1.4节，相关的下载链接也一并附上。本书所有例子都是在NetBeans 8.0.2下开发的，读者需要通过工程的属性对话框自行添加相关API JAR文件的链接。

1.1 什么是NLP

NLP的定义为：NLP是一个使用计算机科学、人工智能和形式语言概念对自然语言进行分析与研究的领域。用通俗的话来讲，就是如何用一些方法和工具从类似网页、文档这样的自然语言资源中得到有意义的、有用的信息。

这不仅是一个有意思的学术难题，也充斥着巨大的商业价值，从它在搜索引擎上的应用就可以看出。用户输入的查询语句经过NLP技术处理后才能返回他们想要的结果，现代的搜索引擎在这方面已经非常做得非常成功了。此外，NLP技术在自动救援系统、复杂问答系统（如IBM Watson项目）中也有广泛应用。

当进行处理语言时，会频繁使用词、语法、语义。语法是构成一个有效语句的规则，例如最常见的英语句子结构是"主语+谓语动词+宾语"结构，如"Tim hit the ball"。而不是其他反常语序，类似"Hit ball Tim"。尽管相比于计算机语言，英语的语法很不严格，但语句基本还是会遵守一定的语法规则。

语义是指一个句子所表达的意思。懂英语的人都明白"Tim hit the ball"这句话的意思。但不管是英语还是其他语言，在很多情况下都存在一定的二义性，一个句子的意思可能需要通过上下文才能真正确定。我们会看到很多机器学习技术是如何尝试理解文本的正确含义的。

后续的讨论将会介绍很多语言学的词汇，这些术语一方面可以帮助读者更好地理解自然语言，另一方面也可以在解释大量NLP技术的时候提供一个共同的词汇表。我们将看到文本如何被拆分成独立的元素，而这些元素又如何被分类。

总的来说，NLP技术是用来增强应用程序的，用程序为用户提供更有价值的服务。NLP技术既包含相对简单的应用，也包含最前沿科技。本书将通过具体实例说明一些可以解决特定问题的简单NLP方法，也会介绍应用于更复杂需求的高级库和类。

1.2 为何使用 NLP

NLP 技术应用的方式多种多样，解决的问题也各有不同。以文本分析为例，它既包括用户在网上输入的简单查询语句，也包括需要生成摘要的大型文件。近些年来，非结构化数据数量在飞速增加，它们以博客、微博等社交媒体的形式广泛散布于整个网络，其数量之大已远远超过人类阅读能力的极限。NLP 技术是分析这些数据的最佳手段。

机器学习和文本分析是提高应用价值的常用手段。常见应用领域如下。

- 搜索：识别文本中的特定元素，可以仅仅是寻找文件中的名字，也可以涉及同义词、误拼写单词分析，以找到与原搜索近似的条目。
- 机器翻译：将一种语言翻译成另一种语言。
- 提取摘要：对段落、文章、文档或文档集合自动提取摘要。NLP 在这一方向已经十分成功。
- 命名实体识别（NER）：从文本中提取地点、任务、事物的名称，通常用于和其他 NLP 任务（如查询）的连接。
- 信息分组：主要基于文本数据，可自动创建一系列反映文件内容的类别。比如有些网站会根据用户需要将内容自动分类列在网页一侧。
- 词性标注（POS）：将文本分隔为不同的语法元素（如名词、动词），作为文本后续分析的基础。
- 情感分析：自动分析人们对书、电影或其他产品的情绪和态度，自动获得消费者对产品的反馈。
- 问答系统：IBM 的 Watson 系统是很好的例子，它已经在 Jeopardy 竞赛中胜过人类，当然智能问答不仅限于玩这种综艺游戏，它在医疗等行业已经有很好的应用。
- 语音识别：虽然人的语音很难分析，但 NLP 技术在语音识别领域取得了很多成就。
- 自然语言生成：通过知识数据自动产生文本，可用于自动播报天气信息或总结医疗报告单。

NLP 技术主要基于机器学习方法解决问题，因此基本流程为训练模型，验证模型准确度，并利用这个模型解决实际问题，1.6 节会对该过程进行详述。

1.3 NLP 的难点

由于多种因素的影响，使得 NLP 非常复杂。比如，世界上有几百种自然语言，每一种都有不一样的语法规则。单词很多具有歧义性，需要结合上下文才能确定其含义。本章主要探讨其中更有意义的几个难点。

在字符层面上，需要考虑以下因素。首先要确定文档的字符编码方式，是 ASCII、UTF-8、UTF-16，还是 Latin-1。其次，需要确定文本是否区分大小写，标点符号和数字是否需要特殊处理，有时候还需要考虑字符表情（一些特定的字符组合、字符图像）、超链接、重复标点（... 或---）、文件后缀名以及夹在名字中的点号等。这些大多会出现在文本预处理阶段，1.7 节会进一步进行说明。

NLP 的第一步是将一个句子/文本拆分成一个词语序列，这些词语称为词项，而这种处理技术就称为分词。对于用空格来分隔单词的语言，分词并不难做，但对于像汉语这样词与词之间无间隔的语言，分词是件十分困难的事情。

第二步，给字词和语素标注语义标签，标注它们是哪种类型的单元。语素是文本中最小的语法单位，例如前缀和后缀。在处理单词的时候，我们往往需要考虑同义词、缩写、首字母缩写、拼写习惯等。

第三步，词干化处理，也就是找出单词中的主干部分，例如单词"walking""walked""walks"的词干是"walk"。搜索引擎通常会使用词干法来处理查询语句。

与词干化类似的还有词形还原。这个过程主要确定单词的基本形式，也称为词元。举个例子，单词"operating"的词干是"oper"，但它的词元是"operate"。词形还原是比词干化更精确的一个过程，通常会基于单词表或形态学技巧来得到词元，很多场景下的分析结果都很精确。

一个个词汇的组合就是短语或句子。但句子的确定并不是简单地寻找句子末尾的句点，因为很多其他地方也会出现句点，如"Ms."，又或者数字"12.834"。

在此基础上，还需要进一步理解句子中哪些是名词，哪些是动词。有时人们并不明确单词之间的指代关系。共指消解很好地解决了这个问题，它主要确定一个单词在一个或多个句子中的指代关系。比如下面这个句子：

"The city is large but beautiful. It fills the entire valley."

句子中的"it"指代"city"。然而，当一个单词有多个意思的时候，需要使用词义消歧算法来确定含义，有时候这很困难。比如"John went back home"中"home"指的是一间房子，一个城市还是其他的什么东西？它的具体意思往往只能通过上下文来推断。比如"John went back home. It was situated at the end of a cul-de-sac."

 尽管困难很多，但在多数情况下NLP技术还是可以合理地处理好这些问题的，并提供附加的信息。比如通过推特数据分析消费者情绪，进而为不满意的消费者提供免费的产品。又如对医疗检查报告进行总结，突出显示比较重要的条目。

提取摘要是指对文本内容产生一个简短描述的过程，这些文本内容可以是多个句子、段落、一个或多个文档。其主旨是找到能表达这个内容的语句，或要理解这些内容的主要信息，或获取用户想得到的条目。通常弄清楚内容的上下文是提取摘要的关键。

1.4 NLP工具汇总

NLP工具有很多。一部分可以借助Java SE SDK使用，但功能有限，只能处理简单的问题；另外一部分是开源库，如Apache的OpenNLP和LingPipe，可以用来解决比较复杂的NLP问题。

Java的字符串类，如String、StringBuilder、StringBuffer可以视为低级的NLP工具，这些类包含了最基本的搜索、匹配、文本替换功能。此外，Java也支持正则表达式，这些正则表达式主要用于特殊编码和子串匹配。Java也为使用正则表达式提供一组丰富的方法。

Java也在一定程度上支持分词（将文本分隔为独立的元素）：

- String类中的split方法
- StreamTokenizer类
- StringTokenizer类

此外，就是各种NLP库和API。下面的表格中列出了部分Java语言的NLP API，大多数都是开源的。当然还有一些是商业 API，本书主要关注开源 API。

API	网址
Apertium	http://www.apertium.org/
文本工程通用架构	http://gate.ac.uk/
Learning Based Java	http://cogcomp.cs.illinois.edu/page/software_view/LBJ
LinguaStream	http://www.linguastream.org/
LingPipe	http://alias-i.com/lingpipe/
Mallet	http://mallet.cs.umass.edu/
MontyLingua	http://web.media.mit.edu/~hugo/montylingua/
Apache OpenNLP	http://opennlp.apache.org/
UIMA	http://uima.apache.org/
Stanford Parser	http://nlp.stanford.edu/software

很多 NLP 任务可以组成一个流水线，里面包含一系列 NLP 任务，可以实现某个目标。支持流水线的示例框架有 GATE 和 Apache UIMA 等。

下一节将会进一步研究部分 NLP API，包括这些 API 的整体框架以及每个 API 包含的一些相关链接。

1.4.1　Apache OpenNLP

Apache OpenNLP 不仅具备了大多数常用的 NLP 功能，还包括执行特定任务、训练模型、测试模型等的组件。OpenNLP 的使用步骤一般是先从文件中实例化一个支持当前任务的模型，然后在该模型上执行相关方法来完成任务。

下面的例子中，我们将对一个简单的字符串进行分词。当然，要让代码正确执行，FileNotFoundException 和 IOException 的异常捕获代码需要补全。使用 try-with-resource 模块，借助 en-token.bin 文件打开一个 FileInputStream 实例，其中 en-token.bin 文件包含一个已使用英文文本训练好的分词模型：

```
try (InputStream is = new FileInputStream(
        new File(getModelDir(), "en-token.bin"))){
    // Insert code to tokenize the text
} catch (FileNotFoundException ex) {
    …
} catch (IOException ex) {
    …
}
```

在 try 块中，用这个文件创建一个 TokenizerModel 类实例，再用这个实例创建 Tokenizer 类实例：

```
TokenizerModel model = new TokenizerModel(is);
Tokenizer tokenizer = new TokenizerME(model);
```

然后调用 tokenize 方法，可以对字符串进行分词，该方法返回一个 String 对象数组：

```
String tokens[] = tokenizer.tokenize("He lives at 1511 W."
 + "Randolph.");
```

使用一个 for 循环输出分词结果，使用中括号标记每个词项：

```
for (String a : tokens) {
  System.out.print("[" + a + "] ");
}
System.out.println();
```

执行完后，可以得到如下结果：

[He] [lives] [at] [1511] [W.] [Randolph] [.]

在这个例子中，tokenizer 识别出 W. 是一个缩略词，而最后的句点是标记句子结尾的分隔词项。

本书中的许多示例都将使用 OpenNLP API。OpenNLP 链接如下表所示：

OpenNLP	网址
官网	https://opennlp.apache.org/
文档	https://opennlp.apache.org/documentation.html
Javadoc	http://nlp.stanford.edu/nlp/javadoc/javanlp/index.html
下载	https://opennlp.apache.org/cgi-bin/download.cgi
Wiki	https://cwiki.apache.org/confluence/display/OPENNLP/Index%3bjsessionid=32B408C73729ACCCDD071D9EC354FC54

1.4.2 Stanford NLP

作为 NLP 领域的领导者，Stanford NLP 组开发了很多 NLP 工具，Stanford CoreNLP 是其中之一。此外，他们还开发了其他的工具组，如 Stanford Parser、Stanford POS tagger、Stanford Classifier 等。Stanford 系列工具同时支持中英文和基本 NLP 功能，包括分词和命名实体识别。

所有这些工具都基于 GPL 开源协议发布，但不允许用于商业应用。发布的 API 不论是组织结构，还是对核心的 NLP 功能的支持度，都非常优秀。

Stanford 工具组支持多种分词技术，本书使用其中的 PTBTokenizer 类来说明 NLP 库的用法。构造器的参数包括一个 Reader 对象，一个 LexedTokenFactory<T> 参数，以及一个指定选项设置的字符串。

LexedTokenFactory 是一个由 CoreLabelTokenFactory 和 WordTokenFactory 类实现的接口。前一个类会保留每个词项在字符串中的起止位置，而后一个类仅仅返回词项，不保留任何位置信息，默认选择后一个类。

下面的例子使用 CoreLabelTokenFactory 类。用待处理字符串初始化一个 StringReader 实例，最后的选项设置字符串设为 null。PTBTokenizer 实现了 Iterator 接口，因此可以用 hasNext 和 next 方法来列出得到的词项列表。

```
PTBTokenizer ptb = new PTBTokenizer(
new StringReader("He lives at 1511 W. Randolph."),
new CoreLabelTokenFactory(), null);
while (ptb.hasNext()) {
  System.out.println(ptb.next());
}
```

输出结果如下：

```
He
lives
at
1511
W.
Randolph
.
```

本书频繁使用 Stanford NLP 库。下表为一些 Stanford NLP 相关链接，主要包括各个库的说明文档和下载链接等。

Stanford NLP	网　址
官网	http://nlp.stanford.edu/index.shtml
核心 NLP	http://nlp.stanford.edu/software/corenlp.shtml#Download
解释器	http://nlp.stanford.edu/software/lex-parser.shtml
POS 记录器	http://nlp.stanford.edu/software/tagger.shtml
java-nlp-user 邮件列表	https://mailman.stanford.edu/mailman/listinfo/java-nlp-user

1.4.3 LingPipe

LingPipe 是一组实现了常用 NLP 功能的工具，支持模型训练和测试。这组工具包括完全免费版与需要许可版，其中，免费版也禁止用于商业产品。

还是以 Tokenizer 类进行分词为例说明 LingPipe 的用法。先声明两个列表，一个用于存放词项，另一个用于存放空格：

```
List<String> tokenList = new ArrayList<>();
List<String> whiteList = new ArrayList<>();
```

接下来，声明一个待分词的字符串：

```
String text = "A sample sentence processed \nby \tthe " +
    "LingPipe tokenizer.";
```

然后创建一个 Tokenizer 类的实例。使用基于 Indo-European factory 类的 Tokenizer 类的静态 tokenizer 方法来创建类实例：

```
Tokenizer tokenizer = IndoEuropeanTokenizerFactory.INSTANCE.
tokenizer(text.toCharArray(), 0, text.length());
```

调用 tokenize 方法可以产生上面定义的两个列表：

```
tokenizer.tokenize(tokenList, whiteList);
```

最后，输出分词结果：

```
for(String element : tokenList) {
  System.out.print(element + " ");
}
System.out.println();
```

结果如下：

A sample sentence processed by the LingPipe tokenizer

LingPipe 相关的链接如下表所示：

LingPipe	网　址
官网	http://alias-i.com/lingpipe/index.html
教程	http://alias-i.com/lingpipe/demos/tutorial/read-me.html
JavaDocs	http://alias-i.com/lingpipe/docs/api/index.html
下载	http://alias-i.com/lingpipe/web/install.html
核	http://alias-i.com/lingpipe/web/download.html
模型	http://alias-i.com/lingpipe/web/models.html

1.4.4 GATE

文本工程通用架构（GATE）是英国 Sheffield 大学开发的一组 Java 工具，支持多种语言和 NLP 任务，还可以用于构建 NLP 处理工作流。

GATE 在提供 API 的同时还包括其他一些工具。GATE Developer 是一个文本可视化工具，可以展示文本标注，通过高亮标记来检查文本。GATE Mimir 对大量来源文本进行索引和搜索。GATE Embedded 具有直接向代码中植入 GATE 的功能。用 GATE 来实现 NLP 任务代码量很少。下面是一些与 GATE 相关的链接。

GATE	网址
官网	https://gate.ac.uk/
参考资料	https://gate.ac.uk/documentation.html
JavaDocs	http://jenkins.gate.ac.uk/job/GATE-Nightly/javadoc/
下载	https://gate.ac.uk/download/
Wiki	http://gatewiki.sf.net/

1.4.5 UIMA

结构信息标准组织（OASIS）是信息商业技术领域的一个协会，提出了一套 NLP 工作流框架标准 UIMA，由 Apache UIMA 支持。

UIMA 虽然是关于工作流的标准，但它也对相关的设计模式、数据表达、文本分析的用户角色进行了描述。UIMA 相关的链接如下：

Apache UIMA	网址
官网	https://uima.apache.org/
参考资料	https://uima.apache.org/documentation.html
JavaDocs	https://uima.apache.org/d/uimaj-2.6.0/apidocs/index.html
下载	https://uima.apache.org/downloads.cgi
Wiki	https://cwiki.apache.org/confluence/display/UIMA/Index

1.5 文本处理概览

NLP 任务种类很多，本书仅对其中一部分进行介绍。下面是本书涉及内容的纲要，在后续章节将一一讨论。

- 文本分词
- 文本断句
- 人物识别
- 词性判断
- 文本分类
- 关系提取
- 方法组合

当然，为实现一些自然语言处理目标，这些任务往往会组合起来使用，这一点在后面章节中会有更多体现。比如分词技术只是一个基础性步骤，它通常会作为某些任务的第一步。

1.5.1 文本分词

文本可以分解成不同类型的基本元素，如词、句子、段落，这些元素的分类方法也有很多。本书的文本分解特指将文本分隔成词，也称为词项。形态学是研究词语结构的专业，我们会用到很多形态学的专业词汇来解释 NLP 技术。词的类型很多，下面列出一些常见的类别：

- **简单词语**：像这句话里这些词都可以算是简单词语。
- **词素**：一个单词中最小的有意义单元成为词素。比如单词"bounded"中，"bound"是词素，词素也包括后缀"ed"。
- **前缀/后缀**：是指词根前后的辅助部分。比如单词"graduation"中"ation"是词根"graduate"的后缀。
- **同义词**：是指具有相同意义的词语。"small"和"tiny"可以算是同义词。解决这一问题需要进行词义消歧。
- **缩写词**：比如我们将 Mister Smith 缩写成 Mr. Smith。
- **首字母缩写**：在计算机等很多专业领域，首字母缩写都十分常见。通常把一些较长词句的首字母组合起来，形成缩写。比如将 FORmula TRANslation 写成 FORTRAN，甚至还有递归缩写 GNU（Gnu's Not Unix），当然还有我们正在讨论的 NLP。

- **缩约词**：像"We'll""won't"这种将两个单词合并的写法，也是很常见的。
- **数字**：普通的数字只包含基本的 0～9 十个字符，但复杂的数字可能还包括小数点、正负号等多种科学计数标志。

在 NLP 任务中，确定这些词项的类型十分重要，比如在断句时，我们需要将句子分隔成词，并判断词是否是句子的结尾。

这一过程称为分词。分词的结果是一组词项，其中决定文本在哪里断开的元素叫作分隔符。对于大多数英文文本来说，空格是常用的分隔符，包括空白符、tab 符、回车符。

分词可以简单，也可以复杂。我们用 String 类的 split 函数来做一个简单的分词。首先定义一个字符串，包含如下待分词文本：

```
String text = "Mr. Smith went to 123 Washington avenue.";
```

split 方法用一个正则表达式参数来指定文本分隔的方式。下面代码中参数是 \\s+，表示一个以上的空格作为分隔的分隔符：

```
String tokens[] = text.split("\\s+");
```

再用一个 for-each 循环显示分词结果：

```
for(String token : tokens) {
  System.out.println(token);
}
```

结果如下所示：

```
Mr.
Smith
went
to
123
Washington
avenue.
```

第 2 章会更深入地探讨分词技术。

1.5.2 文本断句

我们普遍认为断句非常简单，因为在英文中，只需要找到点号、问号、感叹号等这些句终字符就可以确定一个句子的结尾。然而，事情并非想象得这么简单，短语中嵌入的点号("Dr.Smith"或"204 SW. Park Street")等其他因素使得准确断句变难,第 3 章中我们会详细分析。这一步骤也称为句子边界消歧(SBD)。英文的 SBD 相较于句终符明确的汉语、日语更有研究意义。

断句是很多其他 NLP 任务的前导工作,像词性标注、实体检测都需要在独立的句子上进行,问答应用也需要对句子进行识别。要完成这项任务,必须先准确地完成断句。

下面我们用 Stanford 的 DocumentPreprocessor 类来演示如何断句。这个类接受一个简单的文本或 XML 文档,返回一个句子列表,并给出了 Iterable 接口,方便循环遍历。

首先声明一个字符串存放文本:

```
String paragraph = "The first sentence. The second sentence.";
```

对这个字符串创建一个 StringReader 对象,该支持 DocumentPreprocessor 类的构造器所需的 read 方法:

```
Reader reader = new StringReader(paragraph);
DocumentPreprocessor documentPreprocessor =
new DocumentPreprocessor(reader);
```

现在,DocumentPreprocessor 对象里存储了对应的文本段落。我们可以再创建一个字符串列表来存储从文本中提取出来的句子:

```
List<String> sentenceList = new LinkedList<String>();
```

对 documentPreprocessor 的每一个元素进行处理。这些元素是 HasWord 列表,每个 HasWord 对象表示一个词。我们用一个 StringBuilder 来存储 HasWord 列表中的每个元素,然后把它转成字符串加入到上面定义的 sentenceList:

```
for (List<HasWord> element : documentPreprocessor) {
  StringBuilder sentence = new StringBuilder();
  List<HasWord> hasWordList = element;
  for (HasWord token : hasWordList) {
      sentence.append(token).append(" ");
  }
  sentenceList.add(sentence.toString());
}
```

最后,用 for-each 语句输出 sentenceList 中所有的句子:

```
for (String sentence : sentenceList) {
  System.out.println(sentence);
}
```

输出结果如下:

```
The first sentence .
The second sentence .
```

在第 3 章会更深入地介绍 SBD。

1.5.3 人物识别

搜索引擎极大地满足了大多数用户寻找商家网址、电影上映时间等需求,文本处理器可以很轻松地从文档中找到指定的单词或短语。然而如果想进一步考虑同义词的问题,事情就会复杂得多,如何找到与用户搜索词意思相近或相同主题的内容是一个很有难度的任务。

举个例子,比如我们想买一台新的笔记本电脑的时候去网上搜索,当在搜索引擎中搜索指定配置的笔记本时,搜索引擎如何得到我们想要的结果?实际上,搜索引擎通常是在我们进行搜索之前就一直在对商家提供的各种信息进行分析,需要从网上各种杂乱无章的信息中获取到有用的信息,并反馈给用户。

最终展现给用户的结果往往是按照类别进行分组的,通常会将类别显示在网页左侧。比如,笔记本电脑的类别可能包括超级笔记本(Ultrabook)、谷歌笔记本(Chromebook),或者按硬盘大小进行分类。下面是亚马逊搜索结果的一个截图:

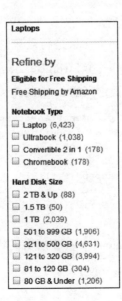

有的搜索十分简单。比如，String 类和其他类似的类都包含 indexOf 和 lastIndexOf 方法，能够寻找字符串在 String 对象中出现的位置。下面的代码就通过 indexOf 方法返回了目标字符串在原来文本中首次出现的位置索引：

```
String text = "Mr. Smith went to 123 Washington avenue.";
String target = "Washington";
int index = text.indexOf(target);
System.out.println(index);
```

输出结果是 22。

当然，这是最简单的搜索问题。

文本检索中，有一个常用的技术叫作反向索引。就是事先对文档进行分词和定位，将比较重要的词和它们出现的位置存储在反向索引中。当用户对文档进行检索的时候，只需要在反向索引中查找到单词，就可以得到它的位置信息。这比每次检索都进行整个文档搜索要快得多。这种数据结构在数据库、信息检索系统和搜索引擎中应用广泛。

更复杂的搜索可能是像"北京哪里的烤鸭店比较好吃？"这样的检索需求。为了回答好这样的提问，需要对查询语句进行实体识别，标注出句子中有意义的词汇，然后对句子进行语义分析，得到句子的意义，最后进行搜索并对候选结果排序。

下面通过组合 tokenizer 和 OpenNLP 中的 TokenNameFinderModel 类来展示寻找名字的过程。由于方法可能会抛出 IOException 的异常，我们用一个 try-catch 来处理，下面代码首先定义了一个存放句子的字符串：

```
try {
    String[] sentences = {
        "Tim was a good neighbor. Perhaps not as good a Bob " +
        "Haywood, but still pretty good. Of course Mr. Adam " +
        "took the cake!"};
    // Insert code to find the names here
} catch (IOException ex) {
    ex.printStackTrace();
}
```

在处理句子前，需要先进行文本分词。下面的代码初始化了一个 Tokenizer 类实例：

```
Tokenizer tokenizer = SimpleTokenizer.INSTANCE;
```

我们需要使用一个模型来断句，这样可以避免将不同句子中的词组合成一个词。我们从 en-ner-person.bin 文件中初始化一个 TokenNameFinderModel 实例：

```
TokenNameFinderModel model = new TokenNameFinderModel(
    new File("C:\\OpenNLP Models", "en-ner-person.bin"));
```

接下来，我们用NameFinderME类来完成寻找名字的任务。我们用TokenName-FinderModel的实例来初始化这个NameFinderME实例：

```
NameFinderME finder = new NameFinderME(model);
```

下面用一个for-each循环来处理多个句子。对每个句子，使用tokenize方法分词，find方法得到一个Span对象数组。这些Span对象存储了find方法识别出的词项起止位置索引：

```
for (String sentence : sentences) {
    String[] tokens = tokenizer.tokenize(sentence);
    Span[] nameSpans = finder.find(tokens);
    System.out.println(Arrays.toString(
        Span.spansToStrings(nameSpans, tokens)));
}
```

输出结果如下：

[Tim, Bob Haywood, Adam]

第4章会对NER进行更深入的讨论。

1.5.4　词性判断

文本分析的另一个任务是词性判断，也就是将一个句子分解成词或词组后，对它们进行分类（名词、动词、形容词、介词等）。我们通常在读小学时就学会怎么判断词性，也知道不要在句子末尾用介词连接下一个句子。

词性检测在提取关系，确定句子意义等NLP任务中有很大作用，确定这些关系称为解析。词性标注可以使数据质量更好，使流水线下游的工作处理起来更容易。

词性标注是很复杂的任务，不过好在这些功能已封装在相关工具和类内，我们可以用OpenNLP的一些类来简单了解词性标注。我们需要一个模型来判断词性，下面代码从en-pos-maxent.bin文件中初始化一个POSModel类实例：

```
POSModel model = new POSModelLoader().load(
    new File("../OpenNLP Models/" "en-pos-maxent.bin"));
```

POSTaggerME类是实际进行标注的类，我们用上面的模型来创建它：

```
POSTaggerME tagger = new POSTaggerME(model);
```

接下来，需要定义一个字符串来存放待处理的句子：

```
String sentence = "POS processing is useful for enhancing the "
    + "quality of data sent to other elements of a pipeline.";
```

然后以 whitespace 分词器对文本进行分词：

```
String tokens[] = WhitespaceTokenizer.INSTANCE.tokenize(sentence);
```

接下来就可以用 POSTTaggerME 类的 tag 方法来进行标注了，将结果存放在一个字符串数组中：

```
String[] tags = tagger.tag(tokens);
```

最后，将标注结果输出：

```
for(int i=0; i<tokens.length; i++) {
    System.out.print(tokens[i] + "[" + tags[i] + "] ");
}
```

输出结果如下：

```
POS[NNP] processing[NN] is[VBZ] useful[JJ] for[IN] enhancing[VBG] the[DT]
quality[NN] of[IN] data[NNS] sent[VBN] to[TO] other[JJ] elements[NNS]
of[IN] a[DT] pipeline.[NN]
```

可以看到每一个词项后面都跟着一个带方括号的缩写词，表示这个词的词性，比如 NNP 表示专有名词。缩写词的具体对应关系，将在第 5 章中完整给出。

1.5.5 文本分类

文本分类是指给文档中找到的信息打上标注，这些标注可能是事先知道的，也可能是不知道的。如果事先给定标注，那么这个过程叫作分类，如果标注未知，我们称之为聚类。

在 NLP 中，分类是比较热点的一个任务，可以把一些文本内容归到某几个组别中。比如军用飞机可以分类为战斗机、轰炸机、侦察机、运输机或救援机。

分类器通常是按照它们输出类型来区分。比如二分分类器，输出结果是 0 或 1，常用于垃圾邮件分类。另外一种为多类分类器，会得到多个可能的类别。

相比于其他 NLP 任务，分类是一个更复杂的流程，其中涉及的步骤我们会在 1.6 节详细讨论，这里就不过多展开了。在第 6 章中，我们会仔细研究分类的过程，并给出一个具体的实例。

1.5.6 关系提取

关系提取是指从文本中找到内在关系,比如下面这个句子"The meaning and purpose of life is plain to see",我们可以知道主题是"The meaning and purpose of life",它与最后一个短语的关系是显而易见的。

人类可以很容易地发现事物之间的关系,特别是表层的关系,而探索事物背后深层次关系则相对困难得多。用计算机去发现文本中所描述事物之间的关系是一个极具挑战性的任务,但好在计算机可以处理庞大的数据集,从海量数据中发现关系,或者在合理的时间内完成关系分析,这是人类无法做到的。

关系有很多种,比如事物的位置,两个人之间的关系,系统的组成部分,谁对某事负责等。关系提取在构建知识库,分析趋势,搜集情报等诸多方面有很大应用价值。通常我们也将关系提取叫作文本分析。

现在,很多技术手段能够实现关系提取,在第 7 章我们会一一介绍。

这里,我们用 Stanford NLP 工具里的 StanfordCoreNLP 类来简单演示一下其中一项技术。该类可以用多个"annotator"组成的流水线来处理文本,一个"annotator"可以认为是一种文本处理操作。我们可以利用 java.util 包中的 Properties 对象给流水线注入多个"annotator"。

首先,我们创建一个 Properties 类实例,然后加入几个 annotator:

```
Properties properties = new Properties();
properties.put("annotators", "tokenize, ssplit, parse");
```

这里我们用了三个 annotator 指定所需操作,第一个 tokenize 是分词,第二个 ssplit 是将分词后的词项分隔组成句子,最后一个 parse 是对句子进行语义分析,解析。

接下来,我们用前面的 properties 实例来初始化一个 StanfordCoreNLP 实例,作为流水线:

```
StanfordCoreNLP pipeline = new StanfordCoreNLP(properties);
```

再接着,用待处理的文本初始化一个 annotation 实例对象:

```
Annotation annotation = new Annotation(
    "The meaning and purpose of life is plain to see.");
```

然后，用流水线对象的 annotate 方法来处理 annotation 对象。最后用 prettyPrint 方法打印出处理结果：

```
pipeline.annotate(annotation);
pipeline.prettyPrint(annotation, System.out);
```

输出结果如下：

```
Sentence #1 (11 tokens):
The meaning and purpose of life is plain to see.
[Text=The CharacterOffsetBegin=0 CharacterOffsetEnd=3 PartOfSpeech=DT]
[Text=meaning CharacterOffsetBegin=4 CharacterOffsetEnd=11
PartOfSpeech=NN] [Text=and CharacterOffsetBegin=12 CharacterOffsetEnd=15
PartOfSpeech=CC] [Text=purpose CharacterOffsetBegin=16
CharacterOffsetEnd=23 PartOfSpeech=NN] [Text=of CharacterOffsetBegin=24
CharacterOffsetEnd=26 PartOfSpeech=IN] [Text=life CharacterOffsetBegin=27
CharacterOffsetEnd=31 PartOfSpeech=NN] [Text=is CharacterOffsetBegin=32
CharacterOffsetEnd=34 PartOfSpeech=VBZ] [Text=plain
CharacterOffsetBegin=35 CharacterOffsetEnd=40 PartOfSpeech=JJ] [Text=to
CharacterOffsetBegin=41 CharacterOffsetEnd=43 PartOfSpeech=TO] [Text=see
CharacterOffsetBegin=44 CharacterOffsetEnd=47 PartOfSpeech=VB] [Text=.
CharacterOffsetBegin=47 CharacterOffsetEnd=48 PartOfSpeech=.]
(ROOT
  (S
    (NP
      (NP (DT The) (NN meaning)
        (CC and)
        (NN purpose))
      (PP (IN of)
        (NP (NN life))))
    (VP (VBZ is)
      (ADJP (JJ plain)
        (S
          (VP (TO to)
            (VP (VB see))))))
    (. .)))

root(ROOT-0, plain-8)
det(meaning-2, The-1)
nsubj(plain-8, meaning-2)
conj_and(meaning-2, purpose-4)
prep_of(meaning-2, life-6)
cop(plain-8, is-7)
aux(see-10, to-9)
```

```
xcomp(plain-8, see-10)
```

输出结果的第一部分是文本分隔的词项和对应的词性标注。之后的树状结构展示了句子的组织结构。最后的部分表示的是各个元素在语法层面的关系。比如下面这一行：

```
prep_of(meaning-2, life-6)
```

它表示单词"meaning"和"life"之间的关系是用介词"of"来表达。这个信息在很多文本简化任务中十分有价值。

1.5.7 方法组合

前面也提到过，NLP 应用问题通常涉及多个基本的 NLP 任务，我们通常将这些组合成一个流水线来得到所需的结果。流水线的使用，在前面 1.5.6 节中已经介绍过。

实际上大多数 NLP 解决方案都会使用流水线。在第 8 章中我们会给出多个例子。

1.6 理解 NLP 模型

虽然解决 NLP 问题的方法和工具种类有很多，但它们之间也会有交集，这一节，我们对这部分内容进行简单介绍。在本书后面的章节中，这些步骤会不断出现，当然，细节上会有少量调整。所以从现在起，理解清楚这些步骤，对于学好 NLP 技术有很大帮助。

这些基本步骤包括：

- 明确目标
- 选择模型
- 构建、训练模型
- 验证模型
- 使用模型

后面部分会对这几个步骤进行讨论。

1.6.1 明确目标

理解清楚待解决的问题十分重要，只有基于对问题的理解，才能很好地设计出一个包含多个 NLP 任务的解决方案。

举个例子，比如我们要回答类似"Who is the mayor of Paris"的问题时，我们需要先将查询语句解析成POS，再确定问题的本质，接着确定问题中符合条件的元素，最后我们从一个创建好的NLP知识库中找到相关信息来回答这个问题。

有的问题并不太复杂。比如说我们可能需要将文本分隔成元素，然后进行分类，类似商家通过产品描述来分析产品的潜在类别。再如通过汽车的描述文本将汽车分成轿车、运动汽车、SUV或小型汽车。

当然，我们必须了解清楚各种已有的NLP任务，这样才能在遇到问题的时候，弄清楚要解决的问题到底是哪一类。

1.6.2 选择模型

我们看到很多NLP任务都是基于模型来做的。比如我们要将一个文档分隔成句子，需要一些算法，然而即使是最好的句子边界判断技术也没法做到百分百断句正确。因此我们通常选择用模型来做文本元素检测，然后根据这些元素的信息来决定该在什么地方断句。

好的模型需要基于所处理的文档类型来构建，比如在历史资料文档中建立的断句模型如果应用到医疗文件中，效果可能会很差。

现在已经有很多开发设计好的模型了，我们可以直接拿来用。一般是根据我们要解决的具体问题，选择一个最合适的模型。但是，当我们找不到合适的模型时，就需要自己去训练新的模型。两种策略的选择往往需要考虑精度和速度的权衡，理解清楚问题的本质和需要的结果质量是选择合适模型的关键。

1.6.3 构建、训练模型

训练模型是对一个数据集执行某个算法，进而调解模型，验证模型的过程。很多时候，我们需要对文本进行全新的处理才能达到预期目标。比如我们在处理新闻文本时候使用的模型，直接用来处理微博数据，效果可能非常差。我们已有的模型往往很难适用于新的数据，当出现这样的情况时，就需要训练新的模型了。

为了训练模型，我们通常需要标记过的数据，也就是知道正确结果的数据。比如做词性标注时，需要使用已标注的文本数据来训练。当模型训练时，我们用这些已有信息来创建、调整模型，这样的数据集一般叫作语料库。

1.6.4 验证模型

一旦模型建好，我们需要用一个样例集对它进行验证。通常还是选择标注数据来验证。我们将模型的预测结果与已知的正确结果进行对比，用来估计模型的精度。通常我们只需要从语料库中选择出一部分数据作为验证数据集，其他作为训练数据集就可以了。

1.6.5 使用模型

使用模型时指将模型应用到实际问题的处理上，具体的应用方式视模型而定。前面的实例中已经有过一些使用方法，比如词性标注部分，我们使用的 POS 模型是从 en-pos-maxent.bin 文件中读取出来的。

1.7 准备数据

搜集和预处理数据是 NLP 中非常重要的一个步骤，包括准备训练数据和处理数据。准备数据有很多因素需要考虑，这里我们主要关注如何用 Java 对文本字符进行处理。

首先，我们需要考虑字符的表达问题。尽管需要处理的主要是英文文本，但有时候也会碰到其他语言的技术问题。这里不光涉及字符编码上的差异，还关系到文本读写顺序的问题。比如日语是从右到左，从上到下的顺序。

编码的类型有多种，包括 ASCII、Latin 以及 Unicode。下表详细列出了编码格式说明，其中 Unicode 比较特殊，是一种较复杂、可扩展的编码方案。

编码	描述
ASCII	采用 128（0～127）个数值对字符进行编码
Latin	包含一些拉丁语的变种（元音变音的不同组合）使用 256 个值编码，很多拉丁语的版本被融合到印欧语系，如土耳其语、世界语
Big5	一种两字节编码方案，用来表示中文字符集
Unicode	有三种 Unicode 方案：UTF-8、UTF-16 和 UTF-32，分别使用 1、2、4 个字节。这种编码可以表示所有现存语言的字符，甚至包括新生的语言，如克林贡语和精灵语

Java 支持以上所有编码方案。我们可以用 javac 程序的 -encoding 命令行选项来指定要使用的编码方案。比如下面命令行指定了用 Big5 编码格式：

```
javac -encoding Big5
```

Java 对字符处理的相关支持，主要用到最基本的数据类型 char、Character 类，以及下表列出的其他一些类和接口：

相关类型	描述
char	基本数据类型
Character	char 类型数据的包装类
CharBuffer	该类支持对一个 char 数组进行 get/put 字符或字符串操作
CharSequence	一个接口，由 CharBuffer、Segment、String、StringBuffer、StringBuilder 等类实现。支持字符串的只读访问

Java 还提供了很多字符串处理的类和接口，下表是一个简单总结。我们在很多例子中都会用到这些类。String、StringBuffer 和 StringBuilder 这些类提供了相似的字符串处理功能，只在可修改性、线程安全性上存在差异。CharaterIterator 接口和 String-CharacterIterator 类提供了遍历字符串的功能。Segment 类表示的是一个文本的一部分。

类／接口	描述
String	不可变的字符串
StringBuffer	表示一个可变的字符串，线程安全
StringBuilder	与 StringBuffer 兼容，但线程不安全
Segment	表示字符数组文本中的一部分，提供字符快速访问方法
CharacterIterator	定义的文本迭代器，支持双向遍历
StringCharacterIterator	实现了 CharacterIterator 的类

此外我们还需要考虑读取文件的格式。通常我们所处理的文字都是有带标注的，比如网页数据，里面的文本是用 HTML 语言标注的。这些不必要的内容，在我们进行处理之前需要删掉。

多目标因特网邮件扩展（MIME）类型是用来区分文件内容格式的，常见的类型如下表所示。我们要么自己写代码去除或修改文件中的标注，要么用其他软件来处理。有一些 NLP 接口提供了相关的格式处理工具。

文件格式	MIME 类型	描述
Text	plain/text	简单文本文件
Office Type Document	application/msword application/vnd.oasis. opendocument.text	微软 Office 文件 Open Office 文件
PDF	application/pdf	Adobe 可移植文件格式
HTML	text/html	网页文件
XML	text/xml	可扩展标记语言
Database	Not applicable	数据库文件可以是很多种形式

很多 NLP 接口会默认输入数据是清洗过的，没有清洗过的数据会使结果出现偏差。

1.8 本章小结

本章我们介绍了 NLP 及其简单应用。可以看到，NLP 可以解决从简单的搜索到复杂的分类等各种各样的问题。本章还通过代码实例展示了 Java 语言对 NLP 的支持，从对字符串的支持，到高级 NLP 库的支持。此外我们还谈到了模型的训练、验证、使用等相关问题。

本书中，我们会讲到如何用简单的和复杂的方法来处理一些基本的 NLP 任务。很多问题用简单的方法就可以处理，在这种情况下，掌握简单技术的使用方法就足够了。当然有的情况较为复杂，需要高级的技术。

下一章，我们将会对文本分词进行深入探索。

CHAPTER 2

第 2 章

文本分词

文本分词即将一段文本分隔成多个独立的单元（词项），以便在这些词项上进行其他处理，包括词干化，词形还原，去除停用词，同义词扩展以及文本转换成小写等。

出于工作需要，我们会介绍一些建立在标准 Java 分布式基础上的分词方法，这种方法不需要导入 NLP 库，然而这些方法也是有限的。随后，我们将讨论 NLP API 支持的具体分词器和分词方法，参考这些例子可以了解分词器如何使用及其输出的具体类型。随后我们对这些方法的异同进行简单总结。

有许多专业化的分词器，比如 Apache Lucene 项目支持多种语言和多种专业文献的分词，WikipediaTokenizer 类分词器专门处理维基百科的文档，ArabicAnalyzer 类专门处理阿拉伯语文本等，不胜枚举。

我们也将检测专业化的分词器如何训练，以便处理专门类型的文本，当遇到不同形式的文本时，可以避免再写一个新的分词器。

接下来，我们将阐述如何应用这些分词器进行词干化，词形还原，去除停用词等具体操作。POS 可认为是一个文本分词的特例。这一话题将在第 5 章中研究。

2.1 理解文本分词

有许多方法可对文本进行分类，比如，我们关心的字符问题，可能需要忽略标点符号或将缩约词展开。而对于单词问题，我们需要进行不同的处理，比如：

- 通过词干化、词形还原识别词素
- 展开缩写词和首字母缩写词
- 分隔开数字单位

由于标点符号有时是单词的一部分，因此并不能总以标点符号来分隔单词，例如单词 can't。还有如何将多个单词组合形成有意义的短语。语句判断也是一个因素。我们不必通过句子边界组合单词。

本章只讨论分词方法和一些相关的技巧，比如词干化，而分词方法在其他 NLP 任务中的使用不在本章内容之列，可以去参考后续章节。

2.2 什么是分词

分词（tokenization）是将一段文本分隔成多个更为简单的单元的过程，一般来说，就是一些单个的单词。词项由一组分隔符分开，分隔符通常是空格，在 Java 中空格被定义为 Character 类的 isWhitespace 方法。下表中列出了这些字符。但有时需要使用不同的分隔符，例如当空格分隔出了难以理解的文本片段时，比如在段落边界，检测到这些文本片段尤为重要。

字符	含义
Unicode space character	（空格分隔符，线分隔符或段落分隔符）
\t	U+0009 横向制表
\n	U+000A 换行
\u000B	U+000B 垂直列表
\f	U+000C 换页
\r	U+000D 回车
\u001C	U+001C 文件分隔符
\u001D	U+001D 群分隔符
\u001E	U+001E 记录分隔符号
\u001F	U+001F 单元分隔符

分词的过程十分复杂，受到以下多种因素影响：

- 语言：不同的语言有其独特的挑战。空格是常用的分隔符，但是对于汉语来说并不适用。
- 文本形式：文本通常以不同格式存储或呈现。相较于简单文本，HTML 或其他标记技术的分词处理过程更为复杂。

- 停用词：对于诸如一般搜索的 NLP 任务来说，常用词可能并不重要，这些常用词被称为停用词，通常将其去掉，例如"a""and""she"等。
- 文本扩展：对于首字母缩写词，有时需要将其展开以便于后期进行文本处理。例如，当搜索关于"机器"的问题时，知道 IBM 是 International Business Machines 的缩写会很有帮助。
- 字母大小写：在某些情况下，单词的大小写十分重要，有助于区分专有名词。进行分词时，将文本转换成大小写一致可以简化搜索。
- 词干化和词形还原：这一过程将单词转换为它们的词根。

去除停用词可以节省索引空间，加速索引过程。然而一些搜索引擎并不去除停用词，因其对于一些特定的查询是有帮助的。例如进行精确匹配时，去除停用词会导致错误。另外，NER 任务往往取决于是否包含停用词。例如"Romeo and Juliet"是部戏剧取决于其间包含停用词"and"。

 有许多规定停用词的列表，但是有时停用词的使用也取决于具体领域。停用词列表可以查看网址 http://www.ranks.nl/stopwords，其中列出了几种英语及其他语言的停用词。http://www.textfixer.com/resources/common-english-words.txt 提供了一个逗号分隔格式的英文停用词列表。

前十位的停用词列表参见下面的表格（来自 Stanford 网站 http://library.stanford.edu/blogs/digital-library-blog/2011/12/stopwords-searchworks-be-ornot-be）。

停用词	出现次数
the	7 578
of	6 582
and	4 106
in	2 298
a	1 137
to	1 033
for	695
on	685
an	289
with	231

以下章节主要关注英文文本分词的方法，通常使用空格或其他分隔符以得到一组词项。

> 解析与分词密切相关，都是识别文本的各个部分，但解析也涉及识别语义及语义间的相互关系。

分词器的使用

分词可以用于拼写检查和简单搜索等任务，对于各种下游 NLP 任务也十分有用，如识别 POS、文本断句和分类等。后面大多数章节中涉及的任务都需要先进行分词处理。

通常，分词只是诸多任务中的一步。这些步骤需要使用流水线技术，这将在本章后面加以说明。分词非常重要，如果分词的结果不理想，那么下游的任务也将受到不利影响。

在 Java 中包含许多种可行的分词器和分词技术。有几个专门设计支持分词的 Java 核心类。其中一些已经过时了。也有一些 NLP API 用来解决既简单又复杂的分词问题。接下来的两节将讨论这些方法。首先，我们将了解 Java 核心类提供的方法，然后将展示一些 NLP 的 API 分词语料库。

2.3 一些简单的 Java 分词器

如下所示是一些支持简单分词的 Java 类：

- Scanner
- String
- BreakIterator
- StreamTokenizer
- StringTokenizer

虽然这些类提供了有限的支持，但了解它们如何使用很有必要。对于某些任务来说，这些类就足够了，没有必要去使用那些难以理解、效率较低的方法。我们将对这些 Java 类分词器一一说明。

StreamTokenizer 和 StringTokenizer 类不应该被用于新的开发。然而 String 类的 split 方法通常是更好的选择。在这里介绍它们是以防你遇到它们会怀疑是否应该使用。

2.3.1 使用 Scanner 类

Scanner 类用于从文本源读取数据，可能是来自标准输入，也可能是来自文件。它提供了简单易用的分词技术。

Scanner 类使用空格作为默认的分隔符。可以使用多个不同的构造器创建 Scanner 类的实例。下面所列的构造器使用了一个简单的字符串，使用 next 方法从输入流取得下一个词项，词项与字符串相隔离，存储在列表中并输出：

```
Scanner scanner = new Scanner("Let's pause, and then "
    + " reflect.");
List<String> list = new ArrayList<>();
while(scanner.hasNext()) {
    String token = scanner.next();
    list.add(token);
}
for(String token : list) {
    System.out.println(token);
}
```

执行后输出结果如下：
```
Let's
pause,
and
then
reflect.
```

这个简单的方法有一些缺点。正如第一个词 Let's，如果需要将其展开并分词，那么这个方法不能完成。此外，这句话的最后一个单词还附带了句号。

指定分隔符

如果我们不满意默认的分隔符，可以使用几种方法来改变它的表现。下表总结了几种方法，可以给你提供一些不错的想法。

方　　法	作　　用
useLocale	使用 locale 设置默认匹配的分隔符
useDelimiter	设置基于字符串或模式的分隔符
useRadix	指定工作时数字的基数
skip	跳过匹配模式的输入并忽略分隔符
findInLine	忽略分隔符找到下一个模式

在这里，我们将展示 useDelimiter 方法的使用。如果在上一节例子中的 while 语句之前直接使用以下语句，分隔符将只能使用空格、单引号和句号。

```
scanner.useDelimiter("[ ,.]");
```

执行后结果如下。使用逗号分隔符使我们得到了一个空白行，也就是它将空字符串作为了一个词项：

Let's
pause

and
then
reflect

此方法使用了字符串中定义的 pattern，括号之间创建了一类字符，是与这三个字符匹配的正则表达式。Java 中 pattern 的解释可以参见 http://docs.oracle.com/javase/8/docs/api/。分隔符列表可以使用 reset 方法复位为空格。

2.3.2 使用 split 方法

我们在第 1 章中阐述了 String 类的 split 方法，具体如下：

```
String text = "Mr. Smith went to 123 Washington avenue.";
String tokens[] = text.split("\\s+");
for (String token : tokens) {
    System.out.println(token);
}
```

输出结果如下：

Mr.
Smith
went
to
123
Washington
avenue.

split 方法也使用了正则表达式。如果我们用上一节使用的字符串（"Let's pause, and then reflect."）替换也将得到相同的结果。

split 方法有一个重载的版本，使用一个整数来制定正则表达式的 pattern 匹配目标文本的次数，当达到该匹配次数时匹配操作停止。

Pattern 类也有 split 方法，它将基于创建 Pattern 对象使用的 pattern 分隔参数。

2.3.3 使用 BreakIterator 类

分词的另一种方法涉及 BreakIterator 类的使用。该类支持不同文本单元的整数边界位置。在本节中，我们将演示如何使用它来分词。

BreakIterator 类有一个默认的受保护的构造器。我们将使用静态的 getWordInstance 方法创建一个类的实例。该方法有一个使用 Locale 对象重载的版本。该类具有多种方法可以获得各种边界，如下表所示。Done 表明已经找到了最后一个边界。

方法	用处
first	返回文本的第一个边界
next	返回目前边界的下一个边界
previous	返回目前边界的上一个边界
setText	将一个字符串与 BreakIterator 实例关联

我们创建一个 BreakIterator 类的实例及一个使用到的字符串：

```
BreakIterator wordIterator = BreakIterator.getWordInstance();
String text = "Let's pause, and then reflect.";
```

将文本分配给实例并定义第一个边界：

```
wordIterator.setText(text);
int boundary = wordIterator.first();
```

下面的循环将使用 begin 和 end 变量存储单词断开的开始和结束边界的索引，边界值是整数，然后将看到每个边界对及其关联文本。

当找到最后的边界，循环终止：

```
while (boundary != BreakIterator.DONE) {
    int begin = boundary;
    System.out.print(boundary + "-");
    boundary = wordIterator.next();
    int end = boundary;
    if(end == BreakIterator.DONE) break;
    System.out.println(boundary + " ["
        + text.substring(begin, end) + "]");
}
```

输出结果如下，括号中明确给出了对应文本：

```
0-5   [Let's]
5-6   [ ]
6-11  [pause]
11-12 [,]
12-13 [ ]
13-16 [and]
16-17 [ ]
17-21 [then]
21-22 [ ]
22-29 [reflect]
29-30 [.]
```

这一方法在识别基础词项的任务上表现极好。

2.3.4　使用 StreamTokenizer 类

java.io 包中的 StreamTokenizer 类用来对输入流文本进行分词，它是一个很早的类，并不像下一节将讨论的 StringTokenizer 类那样灵活。该类的实例通常基于一个文件创建，然后对文件中的文本分词。可以使用字符串构造。

该类使用 nextToken 方法返回下一个词项，返回的词项是一个整数，整数的值反映了返回的词项的类型，可以基于词项类型进行相应处理。

StreamTokenizer 类的成员变量如下表所示：

成员变量	数据类型	含义
nval	double	如果当前词项是一个数字则存有一个数字
sval	String	如果当前词项是一个单词则存有这个词项
TT_EOF	static int	流结束的一个常数
TT_EOL	static int	行结束的一个常数
TT_NUMBER	static int	读取的词项的数量
TT_WORD	static int	指明一个单词词项的常数
ttype	int	读取的词项的类型

在这个例子中，声明 isEOF 变量（这是用来终止循环的）之后创建了一个分词器。nextToken 方法返回词项的类型，基于词项的类型，得到数字和字符串类型的词项：

```
try {
    StreamTokenizer tokenizer = new StreamTokenizer(
        newStringReader("Let's pause, and then reflect."));
    boolean isEOF = false;
    while (!isEOF) {
        int token = tokenizer.nextToken();
        switch (token) {
            case StreamTokenizer.TT_EOF:
                isEOF = true;
                break;
            case StreamTokenizer.TT_EOL:
                break;
            case StreamTokenizer.TT_WORD:
                System.out.println(tokenizer.sval);
                break;
            case StreamTokenizer.TT_NUMBER:
                System.out.println(tokenizer.nval);
                break;
            default:
                System.out.println((char) token);
        }
    }
} catch (IOException ex) {
    // Handle the exception
}
```

执行代码结果如下：

Let

'

这并不是我们想得到的，问题在于分词器使用单引号字符和双引号表示引用的文本。由于没有对应的引号，字符串的其余部分被忽略了。

可以使用 ordinaryChar 方法指定哪些字符应为普通字符，在这里，单引号和逗号字符被指定为普通字符：

```
tokenizer.ordinaryChar('\'');
tokenizer.ordinaryChar(',');
```

加入上述代码后执行结果如下：

Let

'

s

pause

,

and

then

reflect.

单引号问题解决后,这两个字符被视为分隔符并作为词项返回了。还有一种 whitespaceChars 方法可以用来指定哪些字符被视为空格。

2.3.5 使用 StringTokenizer 类

StringTokenizer 类在 java.util 包中,它比 StreamTokenizer 类更灵活,可以处理任何来源的字符串。该类的构造器以被分词的字符串作为参数,然后使用 nextToken 方法返回词项,hasMoreTokens 方法用于判断输入流是否有剩余词项。示例如下:

```
StringTokenizerst = new StringTokenizer("Let's pause, and "
    + "then reflect.");
while (st.hasMoreTokens()) {
    System.out.println(st.nextToken());
}
```

执行结果为:

```
Let's
pause,
and
then
reflect.
```

构造器是重载的,允许指定分隔符以及分隔符是否应该作为一个词项返回。

2.3.6 使用 Java 核心分词法的性能考虑

使用这些 Java 核心分词法时,简要地讨论它们的表现很有必要。由于影响代码执行的各种因素,测量性能有时会很棘手。对几个 Java 核心分词技术的性能比较可以参考 http://stackoverflow.com/questions/5965767/performance-of-stringtokenizer-class-vs-split-method-in-java。对于提到的问题,indexOf 方法是最快的。

2.4 NLP 分词器的 API

这一节将演示几个分别使用 OpenNLP、Stanford 及 LingPipe 的 API 的不同分词方法。尽管还有许多可行的 API,但我们只演示这几个。这些例子将使你了解都有哪些可行的方法。

我们将使用一个名为 paragraph 的字符串来阐述这些方法。字符串包括一个新的换行符,并且可能出现在真实文本的任何一个地方。它在这里定义为:

```
private String paragraph = "Let's pause, \nand then +
    + "reflect.";
```

2.4.1 使用 OpenNLPTokenizer 类分词器

OpenNLP 有一个 Tokenizer 接口,由以下三种类实现:SimpleTokenizer、TokenizerME 和 WhitespaceTokenizer。这个接口支持两种方法:

- tokenize:给定一个字符串进行分词,并返回一组字符串类型的词项。
- tokenizePos:给定一个字符串并返回一组 Span 对象。Span 类用于指定词项的开始和结束的偏移量。

这三个类将在下节介绍。

2.4.1.1 SimpleTokenizer 类的使用

顾名思义,SimpleTokenizer 类用于进行文本的简单分词。如下面的代码所示,INSTANCE 变量用于实例化类。对 paragraph 变量执行 tokenize 方法,可以得到其分词的词项:

```
SimpleTokenizer simpleTokenizer = SimpleTokenizer.INSTANCE;
String tokens[] = simpleTokenizer.tokenize(paragraph);
for(String token : tokens) {
    System.out.println(token);
}
```

执行后结果如下:

Let
'
s
pause
,
and
then
reflect
.

使用这一分词器,标点符号也被作为单独的词项了。

2.4.1.2 WhitespaceTokenizer 类的使用

顾名思义,该类使用空格作为分隔符。在下面的代码中,先是创建了 tokenizer 的实例，然后对输入的 paragraph 执行 tokenize 方法，for 语句用于显示输出结果：

```
String tokens[] =
WhitespaceTokenizer.INSTANCE.tokenize(paragraph);
for (String token : tokens) {
    System.out.println(token);
}
```

执行后结果如下：

Let's

pause,

and

then

reflect.

虽然这并没有将缩约词和类似的文本单元分开，但对于某些应用来说仍是有用的。该类也具有 tokizePos 方法可以得到词项的边界。

2.4.1.3　TokenizerME 类的使用

TokenizerME 类应用了根据最大熵原理（maxent）和一个统计模型创建的模型并用于分词。最大熵模型用于确定数据（或文本）之间的关系。比如来自各种社交媒体的文本并不是格式化的，而且还使用了大量的俚语和特殊符号（如表情）。统计模型的分词器（如最大熵模型）提高了分词结果的质量。

该模型较为复杂，在此不讨论其细节，有兴趣的话可以参考 http://en.wikipedia.org/w/index.php?title=Multinomial_logistic_regression&redirect=no。

TokenizerModel 类隐藏了模型并用于实例化分词器，这个模型一定是预先训练好的。在下面的例子中，使用 en-token.bin 文件中的模型实例化一个分词器，这个模型已经被训练好了，可以应用在普通英文文本上。

模型文件的位置可以通过 getModelDir 方法得到。该方法返回值取决于模型在系统上的存储位置。在 http://opennlp.sourceforge.net/models-1.5/ 这个网站上可以找到许多模型。

在创建一个 FileInputStream 类的实例后，将输入流作为 TokenizerModel 构造器的参数。然后 tokenize 方法将生成一个字符串的数组。接下来的代码用于展示词项：

```
try {
    InputStream modelInputStream = new FileInputStream(
        new File(getModelDir(), "en-token.bin"));
    TokenizerModel model = new
        TokenizerModel(modelInputStream);
    Tokenizer tokenizer = new TokenizerME(model);
    String tokens[] = tokenizer.tokenize(paragraph);
    for (String token : tokens) {
        System.out.println(token);
    }
} catch (IOException ex) {
    // Handle the exception
}
```

所得结果如下：

Let
's
pause
,
and
then
reflect
.

2.4.2 使用 Stanford 分词器

有许多 Stanford NLP API 类支持分词，举例如下：

- PTBTokenizer 类
- DocumentPreprocessor 类
- 流水线的 StanfordCoreNLP 类

后面每个例子都将使用之前定义的 paragraph 字符串。

2.4.2.1 PTBTokenizer 类的使用

这一分词器模仿了 Penn Treebank 3（PTB）分词器（http://www.cis.upenn.edu/~treebank/）。它在其选项和 Unicode 支持方面不同于 PTB。PTBTokenizerr 类支持几个较早的构造器，但是建议使用三参数的构造器。此构造器使用一个 Reader 对象、一个 LexedToken-Factory<T> 参数以及一个指定所需选项的字符串。

LexedTokenFactory 的接口由 CoreLabelTokenFactory 和 WordTokenFactory 类实现。前者可以保留一个词项开始和结束的字符位置，而后者仅仅得到没有任何位置信息的字符串词项。默认使用 WordTokenFactory 类。下面我们将演示这两个类的使用。

CoreLabelTokenFactory 类的使用如下。用 paragraph 创建了一个 StringReader 的实例，最后的参数（这个例子中是 null）是所需选项。Iterator 接口由 PTBTokenizer 类实现，通过 hasNext 和 next 方法展示出每个词项。

```
PTBTokenizer ptb = new PTBTokenizer(
    new StringReader(paragraph), new
CoreLabelTokenFactory(),null);
while (ptb.hasNext()) {
    System.out.println(ptb.next());
}
```

所得结果如下：

Let
's
pause
,
and
then
reflect
.

使用 WordTokenFactory 类可以得到相同结果，如下所示：

```
PTBTokenizerptb = new PTBTokenizer(
    new StringReader(paragraph), new WordTokenFactory(), null);
```

CoreLabelTokenFactory 类的强大功能在于选项参数。这些选项可以控制分词器的行为，比如引号的处理、省略词的补全以及英式英语或美式英语拼写的判断。选项的清单可以参考 http://nlp.stanford.edu/nlp/javadoc/javanlp/edu/stanford/nlp/process/PTBTokenizer.html。

下面的代码通过 CoreLabelTokenFactory 变量 ctf 和一个选项 "invertible=true" 创建了 PTBTokenizer 对象，这一选项可以令我们获得一个 CoreLabel 对象，并通过它获得每个词项首尾的位置：

```
CoreLabelTokenFactory ctf = new CoreLabelTokenFactory();
PTBTokenizer ptb = new PTBTokenizer(
    new StringReader(paragraph),ctf,"invertible=true");
while (ptb.hasNext()) {
    CoreLabel cl = (CoreLabel)ptb.next();
    System.out.println(cl.originalText() + " (" +
        cl.beginPosition() + "-" + cl.endPosition() + ")");
}
```

所得结果如下，括号内的数字表示词项的首尾位置：

```
Let (0-3)
's (3-5)
pause (6-11)
, (11-12)
and (14-17)
then (18-22)
reflect (23-30)
. (30-31)
```

2.4.2.2　DocumentPreprocessor 类的使用

DocumentPreprocessor 类可用于对输入流分词，另外，它实现了 Iterable 接口，更易于遍历分词序列。这一分词器支持简单的文本和 XML 数据的分词。

我们使用 StringReader 类的实例来阐述这一过程：

```
Reader reader = new StringReader(paragraph);
```

然后实例化 DocumentPreprocessor 类得到一个实例：

```
DocumentPreprocessor documentPreprocessor =
    new DocumentPreprocessor(reader);
```

DocumentPreprocessor 类实现了 Iterable<java.util.List<HasWord>> 的接口。HasWord 接口包含两个处理单词的方法：setWord 和 word 方法。后者返回字符串类型的单词。在下面的代码中，DocumentPreprocessor 类将输入文本分隔成句子并存储在列表 List<HasWord> 中。Iterator 对象用于提取一个句子，然后通过 for 语句列出每个词项：

```
Iterator<List<HasWord>> it = documentPreprocessor.iterator();
while (it.hasNext()) {
    List<HasWord> sentence = it.next();
    for (HasWord token : sentence) {
        System.out.println(token);
    }
}
```

执行后结果如下：

```
Let
's
pause
,
and
```

then
reflect
.

2.4.2.3 流水线的使用

我们将用到第 1 章演示过的 StanfordCoreNLP 类，然而，我们先使用一个更为简单的注释字符串对 paragraph 进行分词。如下所示，创建一个 Properties 对象并分配 tokenize 和 ssplit 两个注释。tokenize 注释指定分词过程，ssplit 注释可以分隔句子：

```
Properties properties = new Properties();
properties.put("annotators", "tokenize, ssplit");
```

接下来创建了 StanfordCoreNLP 类和 Annotation 类：

```
StanfordCoreNLP pipeline = new StanfordCoreNLP(properties);
Annotation annotation = new Annotation(paragraph);
```

annotate 方法执行分词过程，然后 prettyPrint 方法列出所有词项：

```
pipeline.annotate(annotation);
pipeline.prettyPrint(annotation, System.out);
```

词项之后列出了各种统计数据，标记了词项首尾的位置信息，如下所示：

```
Sentence #1 (8 tokens):
Let's pause,
and then reflect.
[Text=Let CharacterOffsetBegin=0 CharacterOffsetEnd=3] [Text='s
CharacterOffsetBegin=3 CharacterOffsetEnd=5] [Text=pause
CharacterOffsetBegin=6 CharacterOffsetEnd=11] [Text=,
CharacterOffsetBegin=11 CharacterOffsetEnd=12] [Text=and
CharacterOffsetBegin=14 CharacterOffsetEnd=17] [Text=then
CharacterOffsetBegin=18 CharacterOffsetEnd=22] [Text=reflect
CharacterOffsetBegin=23 CharacterOffsetEnd=30] [Text=.
CharacterOffsetBegin=30 CharacterOffsetEnd=31]
```

2.4.2.4 LingPipe 分词器的使用

LingPipe 支持多种分词器，本节将说明 IndoEuropeanTokenizerFactory 类的使用。后几节将演示 LingPipe 支持的其他分词器。它的 INSTANCE 成员变量提供了 Indo-European 分词器的实例。其中 tokenizer 方法返回基于待处理文本的 Tokenizer 类的实例，如下所示：

```
char text[] = paragraph.toCharArray();
TokenizerFactory tokenizerFactory =
IndoEuropeanTokenizerFactory.INSTANCE;
```

```
Tokenizer tokenizer = tokenizerFactory.tokenizer(text, 0,
text.length);
for (String token : tokenizer) {
    System.out.println(token);
}
```

执行后结果如下：

```
Let
'
s
pause
,
and
then
reflect
.
```

这些分词器支持"正常"文本的分词。在下一节中，我们将展示如何训练一个分词器来应对独特的文本。

2.4.3 训练分词器进行文本分词

当标准分词器处理文本效果不佳时，训练一个分词器十分必要。我们可以创建一个可以用于分词的模型，而不是写一个传统的分词器。

我们需要读取来自一个文件的数据并用它训练一个模型，这些数据存储为一系列由空格和 <SPLIT> 变量分隔的单词。这个 <SPLIT> 变量用于提供识别词项的更多信息，它有助于识别数字之间的中断（如 23.6），以及逗号等标点符号。我们使用的训练集数据存储在文件 training-data.train 中，如下所示：

```
These fields are used to provide further information about how tokens
should be identified<SPLIT>.
They can help identify breaks between numbers<SPLIT>, such as
23.6<SPLIT>, punctuation characters such as commas<SPLIT>.
```

所用数据并不代表特别的文本，但它确实说明了如何注释文本和训练模型的过程。

我们将使用 OpenNLP 中 TokenizerME 类的重载 train 方法创建一个模型，最后的两个参数需要额外解释一下。最大熵模型决定了文本元素之间的关系。

我们可以在此之前指定模型必须处理的特性的数量。这些特征可以当成模型的各个方面。

迭代次数指的是在确定模型参数时，训练过程进行的次数。一些 TokenME 类的参数如下：

参　　数	用　　处
String	所用语言的字符串
ObjectStream<TokenSample>	ObjectStream 参数存有训练数据
boolean	如果为 true，则忽略字母数字数据
int	指定一个特征的处理次数
int	训练最大熵模型的迭代次数

在下面的例子中，我们先定义一个 BufferedOutputStream 对象，用于存储新的模型。示例中使用的几种方法将生成异常，通过 catch 块进行异常处理：

```
BufferedOutputStream modelOutputStream = null;
try {
    …
} catch (UnsupportedEncodingException ex) {
    // Handle the exception
} catch (IOException ex) {
    // Handle the exception
}
```

通过使用 PlainTextByLineStream 类创建 ObjectStream 类的一个实例。它使用训练文件和字符编码方案作为构造器参数，并且用来创建 TokenSample 对象的第二个 ObjectStream 实例。这些对象是包含词项空间信息的文本：

```
ObjectStream<String> lineStream = new PlainTextByLineStream(
    new FileInputStream("training-data.train"), "UTF-8");
ObjectStream<TokenSample> sampleStream =
    new TokenSampleStream(lineStream);
```

现在可以使用下面代码中所示的 train 方法，语言指定为英语，并忽略字母数字信息。特征值和迭代次数分别设置为 5 和 100：

```
TokenizerModel model = TokenizerME.train(
    "en", sampleStream, true, 5, 100);
```

下表中详细给出了 train 方法的参数：

参　　数	含　　义
Language code	指定所用的自然语言的字符串
Samples	示例文本
Alphanumeric optimization	如果是 true，则跳过字母数字数据
Cutoff	一个特征被处理的次数
Iterations	训练模型的迭代次数

接下来的代码将创建一个输出流,然后将模型写进 mymodel.bin 文件。这样模型就可以使用了:

```
BufferedOutputStream modelOutputStream = new BufferedOutputStream(
    new FileOutputStream(new File("mymodel.bin")));
model.serialize(modelOutputStream);
```

这里不讨论输出的详细信息。然而,它基本上记录了训练过程。输出结果最后一部分已被缩略,因为大部分迭代步骤被删除掉了以节省空间,如下所示:

```
Indexing events using cutoff of 5

Dropped event F:[p=2, s=3.6,, p1=2, p1_num, p2=bok, p1f1=23, f1=3, f1_num, f2=., f2_eos, f12=3.]
Dropped event F:[p=23, s=.6,, p1=3, p1_num, p2=2, p2_num, p21=23, p1f1=3., f1=., f1_eos, f2=6, f2_num, f12=.6]
Dropped event F:[p=23., s=6, p1=., p1_eos, p2=3, p2_num, p21=3., p1f1=.6, f1=6, f1_num, f2=,, f12=6,]
  Computing event counts...  done. 27 events
  Indexing...  done.
Sorting and merging events... done. Reduced 23 events to 4.
Done indexing.
Incorporating indexed data for training...
done.
  Number of Event Tokens: 4
      Number of Outcomes: 2
    Number of Predicates: 4
...done.
Computing model parameters ...
Performing 100 iterations.
  1:  ...loglikelihood=-15.942385152878742    0.8695652173913043
  2:  ...loglikelihood=-9.223608340603953    0.8695652173913043
  3:  ...loglikelihood=-8.222154969329086    0.8695652173913043
  4:  ...loglikelihood=-7.885816898591612    0.8695652173913043
  5:  ...loglikelihood=-7.674336804488621    0.8695652173913043
  6:  ...loglikelihood=-7.494512270303332    0.8695652173913043
Dropped event T:[p=23.6, s=,, p1=6, p1_num, p2=., p2_eos, p21=.6, p1f1=6,, f1=,, f2=bok]
  7:  ...loglikelihood=-7.327098298508153    0.8695652173913043
  8:  ...loglikelihood=-7.1676028756216965   0.8695652173913043
  9:  ...loglikelihood=-7.014728408489079    0.8695652173913043
...
100:  ...loglikelihood=-2.3177060257465376    1.0
```

我们可以使用如下所示的模型,这和我们在 2.4.1.3 节中所用方法相同。唯一的区别是所用模型不同：

```
try {
    paragraph = "A demonstration of how to train a
tokenizer.";
    InputStream modelIn = new FileInputStream(new File(
        ".", "mymodel.bin"));
    TokenizerModel model = new TokenizerModel(modelIn);
    Tokenizer tokenizer = new TokenizerME(model);
    String tokens[] = tokenizer.tokenize(paragraph);
    for (String token : tokens) {
        System.out.println(token);
    }
} catch (IOException ex) {
    ex.printStackTrace();
}
```

结果如下：

```
A
demonstration
of
how
to
train
a
tokenizer
.
```

2.4.4 分词器的比较

下面对几种 NLP API 的分词器进行简要比较。下表列出了不同分词器对同一段文本（"Let's pause, \nand then reflect."）生成的词项。此外需注意的是，输出结果基于对类的简单使用，示例中可能不包含影响词项生成情况的选项参数，其目的只是简单地演示基于示例代码和数据的预期输出类型。

简单分词器	空格分词器	Tokenizer ME	PTB 分词器	文本处理器	IndoEuropean TokenizerFactory
Let	Let's	Let	Let	Let	Let
'	pause,	's	's	's	'
s	and	pause	pause	pause	s
pause	then	,	,	,	pause
,	reflect.	and	and	and	,
and		then	then	then	and
then		reflect	reflect	reflect	then
reflect		.	.	.	reflect
					.

2.5 理解标准化处理

标准化处理是将一组单词转换为更统一的序列的过程，这对于文本的后续处理非常有用。将一组单词转换成标准格式，下游其他过程才能够更好处理数据，而不必再处理一些疑难问题。例如，将所有单词转换为小写会简化搜索过程。

标准化处理可以改善文本匹配的结果，比如说"modem router"有多种表达方法，"modem and router""modem & router""modem/router""modem-router"都是一个意思。通过将这些词标准化为常见形式，更容易为购物者提供正确的信息。

然而，标准化处理也有不利的一面。当字母大小写很重要时，将其转换为小写字母会降低搜索的可靠性。

标准化处理包括以下几方面：

- 将字母转变为小写
- 将缩写词展开
- 去除停用词
- 词干化和词形还原

除了缩写词展开外，其他方法都将一一说明，缩写词展开类似于去除停用词，仅仅是把缩写词用其展开式代替。

2.5.1 转换为小写字母

将文本转变为小写是改善搜索结果的一个较为简单的方法。我们可以使用Java内String类的toLowerCase方法，或者使用一些NLP的API，比如LingPipe的LowerCaseTokenizerFactory类。toLowerCase方法演示如下：

```
String text = "A Sample string with acronyms, IBM, and UPPER "
  + "and lowercase letters.";
String result = text.toLowerCase();
System.out.println(result);
```

结果如下：

```
a sample string with acronyms, ibm, and upper and lowercase letters.
```

LingPipe的LowerCaseTokenizerFactory类将在本章后面的2.5.5节进行说明。

2.5.2 去除停用词

去除停用词有许多方法,一个简单的方法是创建一个类来保存和删除停用词;另外,一些 NLP 的 API 提供去除停用词的支持。第一种方法,我们将创建一个称为 StopWords 的简单的类。第二种方法,我们将使用 LingPipe 的 EnglishStopTokenizerFactory 类来演示。

2.5.2.1 创建一个 StopWords 类

去除停用词的过程包括检查词项一组,与一个停用词的列表进行比较,然后从中去除停用词。为了演示这种方法,我们按下表所示创建一个能够完成基本操作的简单类:

构造器 / 方法	用　　处
Default constructor	使用停用词的默认集合
Single argument constructor	使用存储在文件中的停用词
addStopWord	将一个新的停用词添加到内部列表中
removeStopWords	传入一组单词并返回去除停用词后的新的一组单词

如下代码所示,创建一个 StopWords 类,声明了两个实例变量。其中 defaultStop-Words 变量中存储了默认的停用词列表,HashSet 变量 stopwords 列表用于存储处理过程中的停用词:

```java
public class StopWords {

    private String[] defaultStopWords = {"i", "a", "about", "an",
      "are", "as", "at", "be", "by", "com", "for", "from", "how",
      "in", "is", "it", "of", "on", "or", "that", "the", "this",
      "to", "was", "what", "when", "where", "who", "will", "with"};

    private static HashSet stopWords  = new HashSet();
    ...
}
```

之后是类的两个构造器:

```java
public StopWords() {
    stopWords.addAll(Arrays.asList(defaultStopWords));
}

public StopWords(String fileName) {
    try {
        BufferedReader bufferedreader =
                new BufferedReader(new FileReader(fileName));
        while (bufferedreader.ready()) {
```

```
            stopWords.add(bufferedreader.readLine());
        }
    } catch (IOException ex) {
        ex.printStackTrace();
    }
}
```

使用 addStopWord 方法可以很方便地增加一个单词：

```
public void addStopWord(String word) {
    stopWords.add(word);
}
```

removestopwords 方法用于去除停用词，它创建一个 ArrayList 来存储原始单词并传递给方法。for 循环用来从这个列表中去除停用词。contains 方法用于确定提交的单词是否是停用词，是则删除。ArrayList 用于转换为一列字符串并返回。如下所示：

```
public String[] removeStopWords(String[] words) {
    ArrayList<String> tokens =
        new ArrayList<String>(Arrays.asList(words));
    for (int i = 0; i < tokens.size(); i++) {
        if (stopWords.contains(tokens.get(i))) {
            tokens.remove(i);
        }
    }
    return (String[]) tokens.toArray(
        new String[tokens.size()]);
}
```

下列代码介绍 StopWords 的使用。首先，使用默认构造器声明一个 StopWords 类的实例。然后声明了 OpenNLP 的 SimpleTokenizer 类并定义了示例文本，如下所示：

```
StopWords stopWords = new StopWords();
SimpleTokenizer simpleTokenizer = SimpleTokenizer.INSTANCE;
paragraph = "A simple approach is to create a class "
    + "to hold and remove stopwords.";
```

将示例文本分词并传递给 removeStopWords 方法，得到新的分词结果。

```
String tokens[] = simpleTokenizer.tokenize(paragraph);
String list[] = stopWords.removeStopWords(tokens);
for (String word : list) {
    System.out.println(word);
}
```

执行后结果如下。"A"没有被去除是因为它是大写字母，而该类没有进行大小写转换：

```
A
simple
```

```
approach
create
class
hold
remove
stopwords
.
```

2.5.2.2 使用 LingPipe 去除停用词

LingPipe 的 EnglishStopTokenizerFactory 类可以用于识别、去除停用词,这些停用词在 http://alias-i.com/lingpipe/docs/api/com/aliasi/tokenizer/EnglishStopTokenizerFactory.html 中可以找到,包括 a、was、but、he、for 等。

factory 类的构造器需要 TokenizerFactory 实例作为参数。使用 factory 类的 tokenizer 方法进行分词并去除停用词。首先我们声明一个待分词的字符串:

```
String paragraph = "A simple approach is to create a class "
    + "to hold and remove stopwords.";
```

接下来,基于 IndoEuropeanTokenizerFactory 类创建一个 TokenizerFactory 的实例。然后使用 factory 作为参数创建 EnglishStopTokenizerFactory 的实例:

```
TokenizerFactory factory =
IndoEuropeanTokenizerFactory.INSTANCE;
factory = new EnglishStopTokenizerFactory(factory);
```

使用 LingPipe 的 Tokenizer 类和 factory 的 tokenizer 方法对 paragraph 变量中的文本进行处理。tokenizer 方法需要一个字符数组、一个起始位置索引及字符数组长度作为参数。

```
Tokenizer tokenizer = factory.tokenizer(paragraph.toCharArray(),
    0, paragraph.length());
```

下面的 for 语句将遍历处理后的列表:

```
for (String token : tokenizer) {
    System.out.println(token);
}
```

结果如下:

```
A
simple
approach
create
```

```
class
hold
remove
stopwords
.
```

可以看到字母"A"是个停用词,但是没有从词项中删除。这是因为停用词列表使用小写的"a"而不是大写的"A"。结果它漏掉了这个词。我们将在本章后面2.5.5节中修正这个问题。

2.5.3 词干化

找到一个词的词干需要去掉所有前缀或后缀,其剩下的部分则被认为是词干。识别词干有助于发现相似的文本。例如,一个搜索任务可能会寻找像"book"这样的单词的出现,有许多单词包含它,比如 books、booked、bookings 和 bookmark。识别出词干,然后查找它在文档中是否出现往往更有效,在许多情况下,这样都可以改善搜索结果的质量。

词干分析器可能产生一个不是单词的词干。例如,它可能使 bounties、bounty 和 bountiful 等单词具有相同的词干"bounti",这对搜索仍然有用。

词形还原与词干化相似。这一过程是得到其词元(可在字典中找到的),它对于一些搜索同样有帮助。词干化通常被视为一个较为简单的方法,仅仅希望通过去掉一个词项的首尾部分得到其词根"root"。

而词形还原可以被认为是一个较复杂的方法,目标是得到一个词项的形态学或词汇学上的意义。比如说,"having"的词干是"hav"而词元是"have","was"和"been"词干不同,但是词元均是"be"。

词形还原可以比词干化使用更多的计算资源。它们各有其用,取决于需要解决的问题。

2.5.3.1 Porter Stemmer 的使用

Porter Stemmer 是英文常用的词干分析器,其官网主页为 http://tartarus.org/martin/PorterStemmer/,它通过五个步骤得到一个单词的词干。

Apache OpenNLP 1.5.3 中不包含 PorterStemmer 类，其源代码可从 https://svn.apache.org/repos/asf/opennlp/trunk/opennlp-tools/src/main/java/opennlp/tools/stemmer/PorterStemmer.java 处下载，并添加到你的项目中。

下面的例子中我们将向你演示 PorterStemmer 类，其输入一组单词，通过创建 PorterStemmer 类的实例及 stem 方法对每个单词进行词干化处理：

```
String words[] = {"bank", "banking", "banks", "banker", "banked",
    "bankart"};
PorterStemmer ps = new PorterStemmer();
for(String word : words) {
    String stem = ps.stem(word);
    System.out.println("Word: " + word + "  Stem: " + stem);
}
```

执行结果如下：

```
Word: bank   Stem: bank
Word: banking   Stem: bank
Word: banks   Stem: bank
Word: banker   Stem: banker
Word: banked   Stem: bank
Word: bankart   Stem: bankart
```

Bankart 一词通常与"lesion"合用，"Bankart lesion"是指肩膀的一种损伤，与前面的单词没有多大关系。由此可见，提取词干时只使用普通的词缀。

PorterStemmer 类其他有用的方法可见下表：

方　　法	含　　义
add	将一个 char 添加到当前词干的末尾
stem	无参数情况下调用，如果出现不同的词干将返回 true
reset	将词干生成器复原以使用一个不同的单词

2.5.3.2　用 LingPipe 提取词干

LingPipe 的 PorterStemmerTokenizerFactory 类用于提取词干。在下面的例子中，我们仍使用前一节的单词，IndoEuropeanTokenizerFactory 类用来进行初始分词，然后使用 Porter Stemmer 对其词干化。类的定义如下：

```
TokenizerFactory tokenizerFactory =
    IndoEuropeanTokenizerFactory.INSTANCE;
```

```
TokenizerFactory porterFactory =
    new PorterStemmerTokenizerFactory(tokenizerFactory);
```

接下来定义一个数组存储词干。如下所示，我们仍使用上一节定义的数组 words 分别处理每一个单词，分词后其词干存储在 stem 中，最后列出单词及其词干：

```
String[] stems = new String[words.length];
for (int i = 0; i < words.length; i++) {
    Tokenization tokenizer = new Tokenization(words[i],porterFactory);
    stems = tokenizer.tokens();
    System.out.print("Word: " + words[i]);
    for (String stem : stems) {
        System.out.println("   Stem: " + stem);
    }
}
```

执行后结果如下：

```
Word: bank     Stem: bank
Word: banking  Stem: bank
Word: banks    Stem: bank
Word: banker   Stem: banker
Word: banked   Stem: bank
Word: bankart  Stem: bankart
```

介绍了使用 OpenNLP 和 LingPipe 的 Porter Stemmer 的例子，还应当了解一些其他可行的词干分析器，比如 NGrams 和各种概率—算法混合方法。

2.5.4 词形还原

有许多 NLP 的 API 支持词形还原处理，这一节我们将说明 StanfordCoreNLP 和 OpenNLPLemmatizer 这两个类的使用。词形还原的目标是单词的词元，词元可以理解为单词在字典中的形式，比如"was"的词元是"be"。

2.5.4.1 StanfordLemmatizer 类的使用

这里我们用带流水线的 StanfordCoreNLP 类演示词形还原。首先创建流水线，它带有 lemma 等四个注释，如下所示：

```
StanfordCoreNLP pipeline;
Properties props = new Properties();
props.put("annotators", "tokenize, ssplit, pos, lemma");
pipeline = new StanfordCoreNLP(props);
```

这些必要的注释功能如下：

注释器	操 作
tokenize	分词
ssplit	分句
pos	POS 标注
lemma	词形还原
ner	NER
parse	语法解析
dcoref	共指消解

将 Annotation 构造器传入 paragraph 变量并执行 annotate 方法，代码如下：

```
String paragraph = "Similar to stemming is Lemmatization. "
    +"This is the process of finding its lemma, its form " +
    +"as found in a dictionary.";
Annotation document = new Annotation(paragraph);
pipeline.annotate(document);
```

现在我们需要遍历所有句子及其词项。Annotation 和 CoreMap 类的 get 方法会返回指定类型的值。如果没有指定类型的值，它将返回 null。我们将使用这些类来得到词元的列表。

首先，返回句子的列表并处理每个句子、每个单词以获得其词元。句子及词元列表声明如下：

```
List<CoreMap> sentences =
    document.get(SentencesAnnotation.class);
List<String> lemmas = new LinkedList<>();
```

通过两个 for-each 语句遍历所有句子并存储到词元列表中，之后输出结果：

```
for (CoreMap sentence : sentences) {
    for (CoreLabel word : sentence.get(TokensAnnotation.class)) {
        lemmas.add(word.get(LemmaAnnotation.class));
    }
}

System.out.print("[");
for (String element : lemmas) {
    System.out.print(element + " ");
}
System.out.println("]");
```

所得结果如下：

```
[similar to stem be lemmatization . this be the process of find its lemma
, its form as find in a dictionary . ]
```

与原始文本相比，可见其处理效果很不错：

```
Similar to stemming is Lemmatization. This is the process of finding its
lemma, its form as found in a dictionary.
```

2.5.4.2　OpenNLP 中词形还原的使用

OpenNLP 也支持 JWNLDictionary 类的词形还原，该类的构造器需传入用于识别词根的字典文件的路径作为参数。我们使用的是普林斯顿大学的 WordNet 字典（wordnet.princeton.edu）。实际上字典是存储在目录中的一系列文件，这些文件包含了单词及其词根的列表。比如这一节使用的字典存放在 https://code.google.com/p/xssm/downloads/detail?name=SimilarityUtils.zip&can=2&q= 。

JWNLDictionary 类的 getLemmas 方法传入待处理的单词，以及指定单词的 POS，POS 对于匹配实际的单词类型得到精确的结果十分重要。

下面的代码中，我们创建了一个 JWNLDictionary 类的实例，字典的位置以 \\dict\\ 结尾。代码中还定义了示例文本。构造器还可以通过 try-catch 块处理 IOException 和 JWNLException 异常。

```
try {
    dictionary = new JWNLDictionary("…\\dict\\");
    paragraph = "Eat, drink, and be merry, for life is but a dream";
    …
} catch (IOException | JWNLException ex)
    //
}
```

文本初始化之后添加以下语句。首先，按前文所介绍的方法使用 Whitespace-Tokenizer 类对字符串进行分词。然后，将每个词项及一个 POS 类型的空字符串传给 getLemmas 方法。最后即得到结果：

```
String tokens[] =
    WhitespaceTokenizer.INSTANCE.tokenize(paragraph);
for (String token : tokens) {
    String[] lemmas = dictionary.getLemmas(token, "");
    for (String lemma : lemmas) {
        System.out.println("Token: " + token + "  Lemma: "
            + lemma);
    }
}
```

结果如下：

```
Token: Eat,   Lemma: at
Token: drink, Lemma: drink
Token: be     Lemma: be
Token: life   Lemma: life
Token: is     Lemma: is
Token: is     Lemma: i
Token: a      Lemma: a
Token: dream  Lemma: dream
```

除了词项"is"得到了两种词元外,词形还原过程表现良好。"is"的第二个词元是无效的。这说明使用正确的 POS 对于分词十分重要。我们可以使用一个或以上的 POS 标注作为参数传递给 getLemmas 方法。然而,这又产生了另一个问题:如何确定正确的 POS?这一话题我们将在第 5 章详细讨论。

下表中简要列出了一些 POS 标注,来源于 https://www.ling.upenn.edu/courses/Fall_2003/ling001/penn_treebank_pos.html。完整的列表可以查阅宾夕法尼亚大学(Penn)Treebank Tagset(http://www.comp.leeds.ac.uk/ccalas/tagsets/upenn.html)。

标注	描述
JJ	形容词
NN	单数名词或集合名词
NNS	复数名词
NNP	单数专有名词
NNPS	复数专有名词
POS	所有格结束词
PRP	人称代词
RB	副词
RP	助词
VB	动词,基本形式
VBD	动词,过去式
VBG	动词,动名词或现在分词

2.5.5 使用流水线进行标准化处理

这一节我们将使用流水线综合许多标准化的方法。我们扩展一下使用 LingPipe 去除停用词一节中的例子,添加两个额外的 factory 来标准化文本:LowerCaseTokenizer-Factory 和 PorterStemmerTokenizerFactory。

LowerCaseTokenizerFactory 这个工厂添加在 EnglishStopTokenizerFactory 创建之前,

而 PorterStemmerTokenizerFactory 这个工厂添加在它后面，代码如下所示：

```
paragraph = "A simple approach is to create a class "
    + "to hold and remove stopwords.";
TokenizerFactory factory =
    IndoEuropeanTokenizerFactory.INSTANCE;
factory = new LowerCaseTokenizerFactory(factory);
factory = new EnglishStopTokenizerFactory(factory);
factory = new PorterStemmerTokenizerFactory(factory);
Tokenizer tokenizer =
    factory.tokenizer(paragraph.toCharArray(), 0,
    paragraph.length());
for (String token : tokenizer) {
    System.out.println(token);
}
```

结果如下所示：

simpl
approach
creat
class
hold
remov
stopword
.

我们得到了停用词去除后小写的单词词干。

2.6 本章小结

在这一章中，我们说明了文本分词和标准化处理的各种方法。首先基于 Java 核心类（String 类的 split 方法和 StringTokenizer 类）进行简单的分词。当我们决定弃用 NLP 的 API 类时，这些方法十分有用。

接着我们演示了使用 OpenNLP、Stanford 和 LingPipe 的 API 进行分词，它们的实现方法及应用的选项参数各不相同，最后我们对它们的输出结果进行了简要对比。

标准化处理讨论了小写字母转换、缩写展开、去除停用词、词干化及词形还原等方法，我们说明了如何使用 Java 核心类及 NLP 的 API 实现这些方法。

在下一章，我们将说明使用多种 NLP 的 API 进行文本断句的问题。

CHAPTER 3

第 **3** 章

文本断句

文本断句也称作语句边界消歧（Sentence Boundary Disambiguation，SBD）。这个过程对于那些需要对句子进行分析的下游 NLP 任务是非常有用的。例如像词性（POS）和短语分析这样的一些关于句子的典型工作。

在这一章中，我们将进一步深入了解 SBD 的难点。然后，我们将讨论一些能够在某些情况下适用的核心 Java 方法，并且使用各种各样的 NLP API 提供的一些模型。我们同时也会对关于语句检测模型的训练和验证方法进行说明。我们能够增加一些额外的规则来提升它的表现，但是这只在一定程度上有用。随后，训练模型将处理一般和特殊的情况。本章最后将讨论这些模型和它们的使用方法。

3.1 SBD 方法

SBD 是依赖于语言的，并且通常不是很明确。断句的常用方法包括使用一些规则或训练一个模型。断句的简单规则示例如下，满足下列条件，则句子判断为结束：

- 文本被一个句号、疑问号、分号或者感叹号终止。
- 句号不是在缩略词或者数字之后。

虽然这对于很多情况都能表现得很好，但是并非全部情况都如此。例如，确定缩略词通常并不容易，省略号也许会被误认为是句号。

大多数的搜索引擎没有考虑 SBD，它们只关心问题的词项及其位置。执行数据提取的词性标注和其他 NLP 任务将频繁处理单独的句子。语句边界检测能够帮助分开那些可能会跨句的词语。比如，看下面的句子：

"The construction process was over. The hill where the house was built was short."

如果我们搜索"over the hill"，我们将无意地得到它。

本章中一些例子均使用下面的文本来演示 SBD，这段文本由 3 个简单句及一个较复杂的句子组成：

```
private static String paragraph = "When determining the end of sentences "
    + "we need to consider several factors. Sentences may end with "
    + "exclamation marks! Or possibly questions marks? Within "
    + "sentences we may find numbers like 3.14159, abbreviations "
    + "such as found in Mr. Smith, and possibly ellipses either "
    + "within a sentence …, or at the end of a sentence…";
```

3.2　SBD 难在何处

将文本分解成多个语句的困难有以下这些因素：

- 标点符号经常有歧义
- 缩略词常常包含句号
- 语句可能使用引号进行相互嵌套
- 对于更专门的文本，比如推特（tweet）和聊天对话，我们也许需要考虑换行和分句的结束

标点符号歧义现象可以通过句号说明清楚，通常它被用来标记语句结束，然而，它也能被使用在其他环境中，包括缩略词、数字、e-mail 地址和省略号。其他的标点字符，比如问号和感叹号，也被用在嵌套的引号或专门的文本（如文件中的代码）中。

使用句号的一些情形：

- 结束一个句子
- 结束一个缩略词
- 缩略词的结尾同时也是句子结尾
- 省略号
- 以省略号结束的句子

- 嵌套在括号或引号中

对于大部分的语句，我们会在句子结尾遇到一个句号，这让它们能够容易地被辨别出来。然而，当它们以一个缩略词结束，那么就比较难以辨别了。以下是含有带句号的缩略词的句子：

"Mr. and Mrs. Smith went to the ball."

下面两个句子以带句号的缩略词结尾：

"He was an agent of the CIA."
"He was an agent of the C.I.A."

在最后一个句子中，缩略词的每个字母都跟着一个句号。虽然这不常见，但是也许会出现，所以我们不能简单地忽略它。

另一个使得 SBD 困难的问题是判断一个词是否是缩略词。我们不能简单地把所有大写字母序列当成缩略词，也许用户不小心使用了大写字母拼写一个单词，或者文本预处理时将所有字符转换为小写字母。同时，也有一些缩略词包含小写和大写字母序列。为了处理缩略词，有时使用一个有效的缩略词表。然而缩略词通常是特定领域相关的。

省略号进一步使得问题更加困难。它们也许作为单个字符（扩展 ASCII 0x85 或者 Unicode（U+2026））或者是以三个连续句号出现。

另外，还存在 Unicode 水平省略号（U + 2026）、垂直省略号（U + 22EE）及其表现形式（U + FE19）。除此之外，还存在 HTML 编码。对于 Java 来说，使用 \uFE19。这些编码的不同说明了在文本分析之前需要进行良好的预处理。

下面两个句子展示省略号可能使用的情形：

"And then there was … one."
"And the list goes on and on and…"

第二个句子以省略号结束。在一些情况下我们可以按照 MLA 手册（http://www.mlahandbook.org/fragment/public_index）的建议，使用括号来区分省略号，将它加在出现在原始文本的省略号左右。如下所示：

"The people [⋯] used various forms of transportation [⋯]"（Young 73）。

我们也会找到嵌套在另一个句子中的子句，比如：

The man said,"That's not right."

尽管感叹号和问号出现的情形较少，它们也会引起其他的问题。除了句末外，感叹号还可能出现在其他地方。一些词语的情形，比如 Yahoo!，作为单词的一部分。另外，多个感叹号被用来表示强调，比如"Best wishes!!"。这将导致识别成多个句子，事实上它们并不存在。

3.3 理解 LingPipe 的 HeuristicSentenceModel 类的 SBD 规则

还有其他的规则可以用来处理 SBD。LingPipe 的 HeuristicSentenceModel 类使用一系列的词项规则处理 SBD。我们在此列出它们，用它们去观察哪些规则有用。

这个类使用三个词项集合和两个标记去协助处理：

- 可能的结束：这是那些能作为句子最后一个词的词集。
- 不可能的倒数第二：这些短语不能作为句子的倒数第二个词。
- 不可能的开始：这个短语集包含那些不能作为句子开始的短语。
- 括号匹配：这个标记意味着一个句子要直到匹配到所有的括号才能终止。
- 强制最后的边界：这个指定了输入流中必须被当作句子终止符的最后一个词项，即使并不是一个可能的结束。

括号匹配包括 () 和 []。然而，如果文本是畸形的，这个规则将失效。默认词集列在下表中：

可能的结束	不可能的倒数第二	不可能的开始
.	任何单个字母	右括号
..	个人和职业头衔、地位等	,
!	逗号、冒号、引号	;
?	常用缩略词	:
"	方位	
"	企业代号	--

（续）

可能的结束	不可能的倒数第二	不可能的开始
).	时间、月份等	---
	美国的政党	%
	美国的州（not ME or IN）	"
	装运条款	
	地址缩写	

虽然 LingPipe 的 HeuristicSentenceModel 类使用了这些规则，但其他 SBD 工具的方法中也可使用它们。

SBD 的启发式方法也许并不总比其他技术精确。然而，它们能够在一个特定领域中使用，并且有速度快和消耗内存少的优点。

3.4 简单的 Java SBD

有时，文本很简单，Java 核心支持就已经足够。有两种方法进行 SBD：使用正则表达式和使用 BreakIterator 类。我们将分别说明这两种方法。

3.4.1 使用正则表达式

正则表达式比较难以理解。简单的表达式通常不是问题，但当它们变得更复杂的时候，可读性就降低了。当尝试使用它去进行 SBD 的时候，这个是正则表达式的一个限制。

我们将提出两个不一样的正则表达式。第一个表达式对于某些问题域来说非常简单。第二个比较复杂，并且有更好的效果。

在这例子中，我们建立了一个正则表达式类去匹配句号、问号和感叹号。String 类的 split 方法被用来把文本分隔成句子。

```
String simple = "[.?!]";
String[] splitString = (paragraph.split(simple));
for (String string : splitString) {
    System.out.println(string);
}
```

输出如下：

```
When determining the end of sentences we need to consider several factors
 Sentences may end with exclamation marks
```

Or possibly questions marks
Within sentences we may find numbers like 3
14159, abbreviations such as found in Mr
 Smith, and possibly ellipses either within a sentence …, or at the end
of a sentence…

split 方法不管句号是数字还是缩略词的一部分，都以此将文本分开了。

后面的第二段文本有比较好的结果。这个例子改自 http://stackoverflow.com/questions/5553410/regular-expression-match-a-sentence，使用了以下的正则表达式的 Pattern 类：

[^.!?\s] [^.!?]*(?:[.!?](?!['"]?\s|$)[^.!?]*)*[.!?]?['"]?(?=\s|$)

下面的代码每行各有注释：

```
Pattern sentencePattern = Pattern.compile(
    "# Match a sentence ending in punctuation or EOS.\n"
  + "[^.!?\\s]    # First char is non-punct, non-ws\n"
  + "[^.!?]*      # Greedily consume up to punctuation.\n"
  + "(?:          # Group for unrolling the loop.\n"
  + "  [.!?]      # (special) inner punctuation ok if\n"
  + "  (?!['\"]?\\s|$)  # not followed by ws or EOS.\n"
  + "  [^.!?]*    # Greedily consume up to punctuation.\n"
  + ")*           # Zero or more (special normal*)\n"
  + "[.!?]?       # Optional ending punctuation.\n"
  + "['\"]?       # Optional closing quote.\n"
  + "(?=\\s|$)",
  Pattern.MULTILINE | Pattern.COMMENTS);
```

另一种表示这个正则表达式的方法即能够用 http://regexper.com/ 所提供的工具生成。如下图所示，这幅图描述了展示了这个正则表达式如何工作：

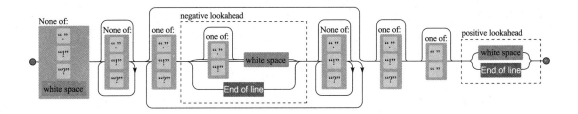

以 paragraph 文本为例执行 matcher 方法，结果展示如下：

```
Matcher matcher = sentencePattern.matcher(paragraph);
while (matcher.find()) {
    System.out.println(matcher.group());
}
```

输出如下，句子的结束符号被保留，但是对于缩略词仍然有很多的问题：

```
When determining the end of sentences we need to consider several
factors.
Sentences may end with exclamation marks!
Or possibly questions marks?
Within sentences we may find numbers like 3.14159, abbreviations such as
found in Mr.
Smith, and possibly ellipses either within a sentence …, or at the end of
a sentence…
```

3.4.2 使用 BreakIterator 类

BreakIterator 类能够用来检测各种各样的文本边界，比如字符、单词、句子和行之间。使用不同的方法创建如下不同的 BreakIterator 类的实例：

- 字符使用 getCharacterInstance 方法
- 单词使用 getWordInstance 方法
- 句子使用 getSentenceInstance 方法
- 行使用 getLineInstance 方法

检测字符的分隔有时是非常重要的，比如，当我们需要处理多个 Unicode 字符（如 ü），这个字符有时由 \u0075(u) 和 \u00a8(¨) 组成。这个类将识别出这些类型的字符。详情可参考 https://docs.oracle.com/javase/tutorial/i18n/text/char.html。

BreakIterator 类可以用于判断句子结尾。它使用一个指针指向当前的边界，提供 next 和 previous 方法使指针在文本中向后和向前移动。BreakIterator 有一个单独的、受保护的默认构造器。为了获得 BreakIterator 的实例来检测句子的结尾，将使用静态（static）方法 getSentenceInstance，如下所示：

```
BreakIterator sentenceIterator =
BreakIterator.getSentenceInstance();
```

同时还有一个重载版本的方法。它使用 Locale 实例作为一个参数：

```
Locale currentLocale = new Locale("en", "US");
BreakIterator sentenceIterator =
    BreakIterator.getSentenceInstance(currentLocale);
```

创建实例的时候，setText 方法将要处理的文本和迭代器（iterator）关联：

```
sentenceIterator.setText(paragraph);
```

BreakIterator 识别了通过各种方法和变量找到的文本边界。所有这些返回整数值，详情可参考下面的表格：

方法	用处
first	返回文本的第一个边界
next	返回当前的边界的下一个边界
previous	返回当前边界的前一个边界
DONE	最后的整数 –1（指示没有找到更多的边界）

为了按顺序使用迭代器，用 first 方法确定第一个边界，重复使用 next 方法获得后续边界。当返回 Done 的时候意味着终止。这一方法使用前面声明的 sentenceIterator 实例，如下所示：

```
int boundary = sentenceIterator.first();
while (boundary != BreakIterator.DONE) {
    int begin = boundary;
    System.out.print(boundary + "-");
    boundary = sentenceIterator.next();
    int end = boundary;
    if (end == BreakIterator.DONE) {
        break;
    }
    System.out.println(boundary + " ["
        + paragraph.substring(begin, end) + "]");
}
```

执行后结果如下：

```
0-75 [When determining the end of sentences we need to consider several
factors. ]
75-117 [Sentences may end with exclamation marks! ]
117-146 [Or possibly questions marks? ]
146-233 [Within sentences we may find numbers like 3.14159 ,
abbreviations such as found in Mr. ]
233-319 [Smith, and possibly ellipses either within a sentence … , or at
the end of a sentence…]
319-
```

这个结果对于简单的句子起作用，但是不能胜任更复杂的情况。

使用正则表达式和 BreakIterator 类都有一定的限制。它们对于那些由简单句子组成的文本效果不错。然而，当文本变得更复杂时，最好使用 NLP 的 API 替代，我们将在下一节介绍。

3.5 使用 NLP API

有许多支持 SBD 的 NLP API 类，其中有些是基于规则的，而其他的模型已经用常见和不常见的文本训练好。我们将介绍利用 OpenNLP、Standford 和 LingPipe 的 API 实现文本断句。

模型也可以被训练以更加完善，相关方法将在 3.6 节进行说明。当处理专门的（如医药或者法律）文本时，需要有一个专门的模型。

3.5.1 使用 OpenNLP

OpenNLP 使用模型进行 SBD。一个 SentDetectorME 类的实例将根据一个模型文件创建，通过 sentDetect 方法返回句子，通过 sentPosDetect 方法返回位置信息。

3.5.1.1 使用 SentenceDetectorME 类

通过 SentenceModel 类从文件中加载模型，然后创建一个 SentenceDetectorME 类的实例，随后调用 sentDetect 方法进行 SBD。这个方法返回一个字符串数组，其中每一个元素是一个句子。

如下面例子所示，用一个 try-with-resources 块来打开包含模型的 en-sent.bin 文件；然后处理字符串 paragraph；接着，（如果有必要）捕获各种 IO 类型异常；最后，使用 for-each 语句列出结果：

```
try (InputStream is = new FileInputStream(
        new File(getModelDir(), "en-sent.bin"))) {
    SentenceModel model = new SentenceModel(is);
    SentenceDetectorME detector = new SentenceDetectorME(model);
    String sentences[] = detector.sentDetect(paragraph);
    for (String sentence : sentences) {
        System.out.println(sentence);
    }
} catch (FileNotFoundException ex) {
    // Handle exception
} catch (IOException ex) {
    // Handle exception
}
```

执行后，我们得到下面的结果：

```
When determining the end of sentences we need to consider several
factors.
Sentences may end with exclamation marks!
Or possibly questions marks?
Within sentences we may find numbers like 3.14159, abbreviations such as
```

```
found in Mr. Smith, and possibly ellipses either within a sentence …, or
at the end of a sentence…
```

可以看到结果还不错。它抓住了简单句和复杂句。当然，它处理文本也并不总是那么完美。下面的段落在一些地方有额外的空格，而一些需要空格的地方缺失空格。这种情况可能发生在聊天会话的分析中：

```
paragraph = " This sentence starts with spaces and ends with "
    + "spaces . This sentence has no spaces between the next "
    + "one.This is the next one.";
```

当用前面的例子处理这一文本时，我们得到结果：

```
This sentence starts with spaces and ends with spaces    .
This sentence has no spaces between the next one.This is the next one.
```

第一个句子的前导空格被删掉，但是结尾的空格没有删除。第三个句子没有被检测出，并且合并到了第二个句子中。

getSentenceProbabilities 方法返回一个 doubles 类型的数组，代表最后使用 sentDetect 文本断句的置信度。在 for-each 语句后增加下面的代码，然后展示结果：

```
double probablities[] = detector.getSentenceProbabilities();
for (double probablity : probablities) {
    System.out.println(probablity);
}
```

所得结果如下：

```
0.9841708738988814
0.908052385070974
0.9130082376342675
1.0
```

这些数字反映 SBD 结果的可靠性极高。

3.5.1.2 使用 setPosDetect 方法

SentenceDetectorME 类的 sentPosDetect 方法返回每个句子的 Span 对象。使用前一节同样的代码，稍作改变，sentDetect 方法替代为 sentPosDetect 方法，for-each 语句如下所示：

```
Span spans[] = sdetector.sentPosDetect(paragraph);
for (Span span : spans) {
    System.out.println(span);
}
```

仍然使用原来的 paragraph 得到结果。Span 对象包含默认的 toString 方法返回的位置信息：

```
[0..74)
[75..116)
[117..145)
[146..317)
```

Span 类内含一些方法。下一段代码演示了 getStart 和 getEnd 方法的使用，可以更清晰地看到 span 对象代表的这些文本：

```
for (Span span : spans) {
    System.out.println(span + "[" + paragraph.substring(
        span.getStart(), span.getEnd()) +"]");
}
```

结果可以看到识别出的句子：

```
 [0..74)[When determining the end of sentences we need to consider
several factors.]
[75..116)[Sentences may end with exclamation marks!]
[117..145)[Or possibly questions marks?]
[146..317)[Within sentences we may find numbers like 3.14159,
abbreviations such as found in Mr. Smith, and possibly ellipses either
within a sentence …, or at the end of a sentence…]
```

还有一些有价值的 Span 方法如下表所示：

方　　法	含　　义
contains	一个重载的方法，确定是否另一个 Span 对象或者索引被包含在目标中
crosses	确定是否两个跨度重叠
length	跨度的长度
startsWith	确定跨度是否由目标跨度作为开头

3.5.2　使用 Stanford API

Stanford 的 NLP 库支持许多文本断句的方法。本节将介绍如何使用下面这些类：

- PTBTokenizer
- DocumentPreprocessor
- StanfordCoreNLP

虽然它们都能进行 SBD，但具体方法各有不同。

3.5.2.1 使用 PTBTokenizer 类

PTBTokenizer 类基于规则进行 SBD，并且有许多分词选项。这个类的构造器需要三个参数：

- 一个封装待处理的文本的 Reader 类
- 一个实现 LexedTokenFactory 接口的对象
- 一个包含分词选项的字符串

这些选项允许我们指定文本、分词器和任何对于某个特定的文本流需要使用的选项。

下面的代码创建了一个 StringReader 类的实例来封装文本，使用了 null 选项的 CoreLabelTokenFactory 类：

```
PTBTokenizer ptb = new PTBTokenizer(new StringReader(paragraph),
    new CoreLabelTokenFactory(), null);
```

我们将使用 WordToSentenceProcessor 类创建一个 List 类的列表来保留句子及其词项。它的 process 方法根据 PTBTokenizer 实例产生的词创建 List 类的列表，如下所示：

```
WordToSentenceProcessor wtsp = new WordToSentenceProcessor();
List<List<CoreLabel>> sents = wtsp.process(ptb.tokenize());
```

这个 List 类的 List 实例能够用几种方法展示出来。下面，List 类的 toString 方法展示了由括号括起来且由逗号分隔的元素的列表：

```
for (List<CoreLabel> sent : sents) {
    System.out.println(sent);
}
```

结果如下：

```
[When, determining, the, end, of, sentences, we, need, to, consider,
several, factors, .]
[Sentences, may, end, with, exclamation, marks, !]
[Or, possibly, questions, marks, ?]
[Within, sentences, we, may, find, numbers, like, 3.14159, ,,
abbreviations, such, as, found, in, Mr., Smith, ,, and, possibly,
ellipses, either, within, a, sentence, ..., ,, or, at, the, end, of, a,
sentence, ...]
```

另一个方法可以将每句话按行列出：

```
for (List<CoreLabel> sent : sents) {
    for (CoreLabel element : sent) {
```

```
        System.out.print(element + " ");
    }
    System.out.println();
}
```

结果如下:

```
When determining the end of sentences we need to consider several factors
.
Sentences may end with exclamation marks !
Or possibly questions marks ?
Within sentences we may find numbers like 3.14159 , abbreviations such as
found in Mr. Smith , and possibly ellipses either within a sentence ... ,
or at the end of a sentence ...
```

如果我们仅仅关心词的位置和句子,可以使用 endPosition 方法,说明如下:

```
for (List<CoreLabel> sent : sents) {
    for (CoreLabel element : sent) {
        System.out.print(element.endPosition() + " ");
    }
    System.out.println();
}
```

执行程序得到下面的结果。每一行的最后一个数字就是句子边界的索引:

```
4 16 20 24 27 37 40 45 48 57 65 73 74
84 88 92 97 109 115 116
119 128 138 144 145
152 162 165 169 174 182 187 195 196 210 215 218 224 227 231 237 238 242
251 260 267 274 276 285 287 288 291 294 298 302 305 307 316 317
```

下列代码可以列出每一个句子的第一个元素及其索引:

```
for (List<CoreLabel> sent : sents) {
    System.out.println(sent.get(0) + " "
        + sent.get(0).beginPosition());
}
```

结果如下:

```
When 0
Sentences 75
Or 117
Within 146
```

如果我们关注句子最后一个元素,可以使用下面的代码。列表中元素表示结束符及其终止位置:

```
for (List<CoreLabel> sent : sents) {
```

```
        int size = sent.size();
        System.out.println(sent.get(size-1) + " "
            + sent.get(size-1).endPosition());
}
```

所得结果如下:

```
. 74
! 116
? 145
... 317
```

调用 PTBTokenizer 类的构造器有许多的可选选项。这些选项被包含在构造器的第三个参数中。选项字符串由逗号隔开的选项组成,比如下面所示:

"americanize=true,normalizeFractions=true,asciiQuotes=true".

一些选项在下表中列出:

选 项	含 义
invertible	用来表明词和空白符必须保留,保证原字符串能够被重建
tokenizeNLs	表明行的结束必须作为一个词
americanize	如果为 true,这将把英国拼写重写为美国拼写
normalizeAmpersandEntity	将 XML& 记号转化为 &
normalizeFractions	将分数字符比如 ½ 转化成长格式(1/2)
asciiQuotes	将引号字符转化成更简单的 ' 或 " 字符
unicodeQuotes	将引号字符转化成 U+2018 到 U+201D 的字符

下面说明了这些选项字符串的使用:

```
paragraph = "The colour of money is green. Common fraction "
    + "characters such as ½ are converted to the long form 1/2. "
    + "Quotes such as "cat" are converted to their simpler form.";
ptb = new PTBTokenizer(
    new StringReader(paragraph), new CoreLabelTokenFactory(),
    "americanize=true,normalizeFractions=true,asciiQuotes=true");
wtsp = new WordToSentenceProcessor();

sents = wtsp.process(ptb.tokenize());
for (List<CoreLabel> sent : sents) {
    for (CoreLabel element : sent) {
        System.out.print(element + " ");
    }
    System.out.println();
}
```

结果如下:

```
The color of money is green .
Common fraction characters such as 1/2 are converted to the long form 1/2
.
Quotes such as " cat " are converted to their simpler form .
```

英式单词拼写 colour 被转化成美式拼写。分数 ½ 被扩展成三个字符：1/2，在最后一个句子中，花引号被转化成它们的简单形式。

3.5.2.2 使用 DocumentPreprocessor 类

当创建 DocumentPreprocessor 类的一个实例，通过传入的 Reader 参数产生一个句子列表。它同样实现了 Iterable 接口，使得遍历这个列表更简单。

在以下的例子中，paragraph 被用来创建一个 StringReader 对象，并且这个对象被用来实例化 DocumentPreprocessor：

```
Reader reader = new StringReader(paragraph);
DocumentPreprocessor dp = new DocumentPreprocessor(reader);
for (List sentence : dp) {
    System.out.println(sentence);
}
```

执行后，我们得到下面的结果：

```
[When, determining, the, end, of, sentences, we, need, to, consider,
several, factors, .]
[Sentences, may, end, with, exclamation, marks, !]
[Or, possibly, questions, marks, ?]
[Within, sentences, we, may, find, numbers, like, 3.14159, ,,
abbreviations, such, as, found, in, Mr., Smith, ,, and, possibly,
ellipses, either, within, a, sentence, ..., ,, or, at, the, end, of, a,
sentence, ...]
```

默认情况下，用 PTBTokenizer 进行分词。setTokenizerFactory 方法能用来指定不同的分词器。下表详细列出了其他可能用到的方法：

方　　法	目　　的
setElementDelimiter	它的参数指定一个 XML 元素，只有在这些元素中的文本才会被处理
setSentenceDelimiter	处理器将假设字符串参数是一个句子的分隔符
setSentenceFinalPuncWords	它的字符串数组参数指定句子结束的分隔符
setKeepEmptySentences	当使用空白符模型时，如果它的参数为 true，那么空的句子将会被保留

这个类能够处理纯文本和 XML 文档。

为了展示如何处理 XML 文件,我们将创建一个简单的 XML 文件 XMLText.xml,文件内容如下:

```
<?xml version="1.0" encoding="UTF-8"?>
<?xml-stylesheet type="text/xsl"?>
<document>
    <sentences>
        <sentence id="1">
            <word>When</word>
            <word>the</word>
            <word>day</word>
            <word>is</word>
            <word>done</word>
            <word>we</word>
            <word>can</word>
            <word>sleep</word>
            <word>.</word>
        </sentence>
        <sentence id="2">
            <word>When</word>
            <word>the</word>
            <word>morning</word>
            <word>comes</word>
            <word>we</word>
            <word>can</word>
            <word>wake</word>
            <word>.</word>
        </sentence>
        <sentence id="3">
            <word>After</word>
            <word>that</word>
            <word>who</word>
            <word>knows</word>
            <word>.</word>
        </sentence>
    </sentences>
</document>
```

我们将再次使用上个例子的代码,将文件替换为 XMLText.xml 文件,使用 DocumentPreprocessor.DocType.XML 作为 DocumentPreprocessor 类构造器的第二个参数,如下所示。这会指定处理器将文本当作 XML 文本处理。另外,我们将指定仅处理 <sentence> 标注中的元素:

```
try {
    Reader reader = new FileReader("XMLText.xml");
    DocumentPreprocessor dp = new DocumentPreprocessor(
        reader, DocumentPreprocessor.DocType.XML);
    dp.setElementDelimiter("sentence");
```

```
        for (List sentence : dp) {
            System.out.println(sentence);
        }
} catch (FileNotFoundException ex) {
    // Handle exception
}
```

这个例子的结果如下：

[When, the, day, is, done, we, can, sleep, .]
[When, the, morning, comes, we, can, wake, .]
[After, that, who, knows, .]

使用 ListIterator 可以使结果更简洁：

```
for (List sentence : dp) {
    ListIterator list = sentence.listIterator();
    while (list.hasNext()) {
        System.out.print(list.next() + " ");
    }
    System.out.println();
}
```

结果如下：

When the day is done we can sleep .
When the morning comes we can wake .
After that who knows .

如果我们没有指定一个元素分隔符，每个单词将呈现如下：

[When]
[the]
[day]
[is]
[done]
...
[who]
[knows]
[.]

3.5.2.3 使用 StanfordCoreNLP 类

StanfordCoreNLP 类用 ssplit 注释支持文本断句。在下面的例子中，使用了 tokenize 和 ssplit 注释。创建了一个流水线对象，并且采用 annotate 方法：

```
Properties properties = new Properties();
properties.put("annotators", "tokenize, ssplit");
```

```
StanfordCoreNLP pipeline = new StanfordCoreNLP(properties);
Annotation annotation = new Annotation(paragraph);
pipeline.annotate(annotation);
```

结果包含很多信息。第一行的结果如下所示:

```
Sentence #1 (13 tokens):
When determining the end of sentences we need to consider several
factors.

[Text=When CharacterOffsetBegin=0 CharacterOffsetEnd=4]
[Text=determining CharacterOffsetBegin=5 CharacterOffsetEnd=16]
[Text=the CharacterOffsetBegin=17 CharacterOffsetEnd=20]
[Text=end CharacterOffsetBegin=21 CharacterOffsetEnd=24] [Text=of
CharacterOffsetBegin=25 CharacterOffsetEnd=27] [Text=sentences
CharacterOffsetBegin=28 CharacterOffsetEnd=37] [Text=we
CharacterOffsetBegin=38 CharacterOffsetEnd=40] [Text=need
CharacterOffsetBegin=41 CharacterOffsetEnd=45] [Text=to
CharacterOffsetBegin=46 CharacterOffsetEnd=48] [Text=consider
CharacterOffsetBegin=49 CharacterOffsetEnd=57] [Text=several
CharacterOffsetBegin=58 CharacterOffsetEnd=65] [Text=factors
CharacterOffsetBegin=66 CharacterOffsetEnd=73] [Text=.
CharacterOffsetBegin=73 CharacterOffsetEnd=74]
```

此外我们还能使用 xmlPrint 方法,它生成 XML 格式的结果,能够比较容易提取感兴趣的信息。这个方法如下所示,它需要我们处理 IOException 异常:

```
try {
    pipeline.xmlPrint(annotation, System.out);
} catch (IOException ex) {
    // Handle exception
}
```

部分结果如下:

```
<?xml version="1.0" encoding="UTF-8"?>
<?xml-stylesheet href="CoreNLP-to-HTML.xsl" type="text/xsl"?>
<root>
  <document>
    <sentences>
      <sentence id="1">
        <tokens>
          <token id="1">
            <word>When</word>
            <CharacterOffsetBegin>0</CharacterOffsetBegin>
            <CharacterOffsetEnd>4</CharacterOffsetEnd>
          </token>
...
          <token id="34">
            <word>...</word>
            <CharacterOffsetBegin>316</CharacterOffsetBegin>
            <CharacterOffsetEnd>317</CharacterOffsetEnd>
          </token>
```

```
        </tokens>
      </sentence>
    </sentences>
  </document>
</root>
```

3.5.3 使用 LingPipe

LingPipe 使用类的层次结构进行 SBD，如下图所示。其底层是 AbstractSentence-Model 类，它主要的方法是一个重载的 boundaryIndices 方法。这个方法返回一个界限索引的整数数组，每个元素代表一个句子边界。

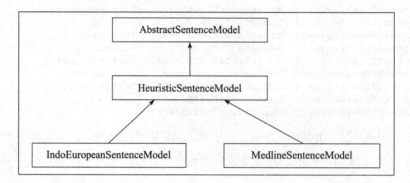

从这个类派生出 HeuristicSentenceModel 类。这个类用到 3.3 节中讨论的可能的结束、不可能的倒数第二和不可能的开始词项。

IndoEuropeanSentenceModel 和 MedlineSentenceModel 类都是从 HeuristicSentenceModel 类派生出，它们已经分别用英语和专业医学文本训练过。下面我们说明一下这两个类。

3.5.3.1 使用 IndoEuropeanSentenceModel 类

IndoEuropeanSentenceModel 类是针对英文文本的。它的构造器带有两个参数，分别指定：

- 最后一个词项是否必为停用词
- 是否需要括号匹配

默认的构造器没有强制最后一个词项必须是停用词或者必须括号匹配。这个句子模型必须和一个分词器一起使用。我们将使用 IndoEuropeanTokenizerFactory 类的默认构造器来达到这个目的，如下所示：

```
TokenizerFactory TOKENIZER_FACTORY=
IndoEuropeanTokenizerFactory.INSTANCE;
SentenceModel sentenceModel = new IndoEuropeanSentenceModel();
```

构造一个分词器,并调用 tokenize 方法来填充这两个列表:

```
List<String> tokenList = new ArrayList<>();
List<String> whiteList = new ArrayList<>();
Tokenizer tokenizer= TOKENIZER_FACTORY.tokenizer(
    paragraph.toCharArray(),0, paragraph.length());
tokenizer.tokenize(tokenList, whiteList);
```

boundaryIndices 方法返回一个整数的边界索引数组。这个方法需要两个包含词项和空格的 String 数组参数。tokenize 方法使用两个 List,这意味着我们需要将 List 转换成等价的数组形式,如下所示:

```
String[] tokens = new String[tokenList.size()];
String[] whites = new String[whiteList.size()];
tokenList.toArray(tokens);
whiteList.toArray(whites);
```

我们可以使用 boundaryIndices 方法来展示列出这些索引:

```
int[] sentenceBoundaries=
sentenceModel.boundaryIndices(tokens, whites);
for(int boundary : sentenceBoundaries) {
    System.out.println(boundary);
}
```

结果如下:

12
19
24

为了显示实际的句子,我们将使用下面的代码。空格索引和词项相差 1:

```
int start = 0;
for(int boundary : sentenceBoundaries) {
    while(start<=boundary) {
        System.out.print(tokenList.get(start)
    + whiteList.get(start+1));
        start++;
    }
    System.out.println();
}
```

结果如下:

When determining the end of sentences we need to consider several factors.

```
Sentences may end with exclamation marks!
Or possibly questions marks?
```

不幸的是，它遗失了最后一个句子。这是因为最后一个句子以省略号结尾。如果我们在后面增加一个句号，我们将得到这样的结果：

```
When determining the end of sentences we need to consider several
factors.
Sentences may end with exclamation marks!
Or possibly questions marks?
Within sentences we may find numbers like 3.14159, abbreviations such as
found in Mr. Smith, and possibly ellipses either within a sentence …, or
at the end of a sentence….
```

3.5.3.2 使用 SentenceChunker 类

一个可行的方法就是使用 SentenceChunker 类进行 SBD。这个类的构造器需要一个 TokenizerFactory 对象和一个 SentenceModel 对象，如下所示：

```
TokenizerFactory tokenizerfactory =
IndoEuropeanTokenizerFactory.INSTANCE;
SentenceModel sentenceModel = new IndoEuropeanSentenceModel();
```

SentenceChunker 实例使用 tokenizer 工厂和句子实例来创建：

```
SentenceChunker sentenceChunker =
    new SentenceChunker(tokenizerfactory, sentenceModel);
```

SentenceChunker 类通过 chunk 方法实现 Chunker 接口。这个方法返回一个实现 Chunking 接口的对象。这个对象通过一个字符序列（CharSequence）指定文本的"语块"。

chunk 方法使用一个字符数组和数组内索引来指定文本的哪些部分需要被处理。返回的 Chunking 对象如下所示：

```
Chunking chunking = sentenceChunker.chunk(
    paragraph.toCharArray(),0, paragraph.length());
```

我们使用 Chunking 对象有两个目的。首先，我们将使用它的 chunkSet 方法返回一个 chunk 对象的集合。然后我们将获得一个包含所有句子的字符串：

```
Set<Chunk> sentences = chunking.chunkSet();
String slice = chunking.charSequence().toString();
```

一个 Chunk 对象存储语句边界的字符偏移量。我们将使用它的 start 和 end 方法结合切片展示句子，如后面展示。每个元素 sentence 保存了语句的边界。我们使用这些信息显示切片中的每一个句子：

```
for (Chunk sentence : sentences) {
    System.out.println("[" + slice.substring(sentence.start(),
        sentence.end()) + "]");
}
```

下面是结果。然而，它在遇到以省略号结束的句子时仍然会出现问题。所以在文本处理之前，在最后一个句子的后面加了一个句号。

```
[When determining the end of sentences we need to consider several
factors.]
[Sentences may end with exclamation marks!]
[Or possibly questions marks?]
[Within sentences we may find numbers like 3.14159, abbreviations such as
found in Mr. Smith, and possibly ellipses either within a sentence …, or
at the end of a sentence….]
```

虽然 IndoEuropeanSentenceModel 类对于英语文本来说还不错，但对于专业文本也许不管用。我们将演示一个训练过的可用于处理医学文本的类（MedlineSentenceModel 类）的使用。

3.5.3.3 使用 MedlineSentenceModel 类

LingPipe 语句模型使用 MEDLINE，它包含很多生物医学文献，以 XML 格式存储，并由美国国家医学图书馆维护（http://www.nlm.nih.gov/）。

LingPine 使用它的 MedlineSentenceModel 类进行 SBD。这个模型已经用 MEDLINE 数据训练过。它使用简单文本，并将它分成词项和空格。MEDLINE 模型可用于文本断句。

在下一个例子中，我们将使用一个从 http://www.ncbi.nlm.nih.gov/pmc/articles/PMC3139422/ 获取的段落来演示这个模型的使用，如下：

```
paragraph = "HepG2 cells were obtained from the American Type
Culture "
    + "Collection (Rockville, MD, USA) and were used only until "
    + "passage 30. They were routinely grown at 37°C in Dulbecco's "
    + "modified Eagle's medium (DMEM) containing 10 % fetal bovine "
    + "serum (FBS), 2 mM glutamine, 1 mM sodium pyruvate, and 25 "
    + "mM glucose (Invitrogen, Carlsbad, CA, USA) in a humidified "
    + "atmosphere containing 5% CO2. For precursor and 13C-sugar "
    + "experiments, tissue culture treated polystyrene 35 mm "
    + "dishes (Corning Inc, Lowell, MA, USA) were seeded with 2 "
    + "× 106 cells and grown to confluency in DMEM.";
```

下面的代码是基于 SentenceChunker 类的，如上一节所示那样，不同之处在于使用了

MEDLINESentenceModel 类：

```
TokenizerFactory tokenizerfactory =
    IndoEuropeanTokenizerFactory.INSTANCE;
MedlineSentenceModel sentenceModel = new
    MedlineSentenceModel();
SentenceChunker sentenceChunker =
    new SentenceChunker(tokenizerfactory,
sentenceModel);
Chunking chunking = sentenceChunker.chunk(
    paragraph.toCharArray(), 0, paragraph.length());
Set<Chunk> sentences = chunking.chunkSet();
String slice = chunking.charSequence().toString();
for (Chunk sentence : sentences) {
    System.out.println("["
        + slice.substring(sentence.start(),
sentence.end())
        + "]");
}
```

结果如下：

```
[HepG2 cells were obtained from the American Type Culture Collection
(Rockville, MD, USA) and were used only until passage 30.]
[They were routinely grown at 37°C in Dulbecco's modified Eagle's medium
(DMEM) containing 10 % fetal bovine serum (FBS), 2 mM glutamine, 1 mM
sodium pyruvate, and 25 mM glucose (Invitrogen, Carlsbad, CA, USA) in a
humidified atmosphere containing 5% CO2.]
[For precursor and 13C-sugar experiments, tissue culture treated
polystyrene 35 mm dishes (Corning Inc, Lowell, MA, USA) were seeded with
2 × 106 cells and grown to confluency in DMEM.]
```

当用医学文本执行时，这个模型将比其他模型有更好的表现。

3.6 训练文本断句模型

我们将使用一个 OpenNLP 的 SentenceDetectorME 类来说明训练过程。此类有一个静态 train 方法，使用文件中的示例语句。该方法返回一个通常序列化为一个文件的模型，供以后使用。

模型使用特别的标注数据来确定句子结束位置。通常，需要一个大文件来提供训练目的的样本。部分文件用来训练，剩下的用来在模型训练后对其进行验证。

OpenNLP 使用的训练文件每行由一个句子组成。通常，至少需要 10 ～ 20 个句子样本来避免错误。为了说明这个过程，我们使用文件 sentence.train，它是《Twenty Thousand Leagues under the Sea》（Jules Verne）一书的第 5 章，该书可从 http://www.gutenberg.org/

files/164/164-h/164-h.htm#chap05 找到。

FileReader 对象用来打开文件，这个对象被用作 PlainTextByLineStream 构造器的参数。输出流包含文件每一行的每一个字符串，它作为 SentenceSampleStream 构造器的参数，将句子的字符串转换成 SentenceSample 对象，这些对象存储每个句子开始的索引。代码如下，语句被包含在一个 try 块中，以便处理异常：

```
try {
    ObjectStream<String> lineStream = new PlainTextByLineStream(
        new FileReader("sentence.train"));
    ObjectStream<SentenceSample> sampleStream
        = new SentenceSampleStream(lineStream);
    ...
} catch (FileNotFoundException ex) {
    // Handle exception
} catch (IOException ex) {
    // Handle exception
}
```

接下来，像下面这样使用 train 方法：

```
SentenceModel model = SentenceDetectorME.train("en",
    sampleStream, true,
    null, TrainingParameters.defaultParams());
```

所得结果是一个训练好的模型。这个方法的参数详情见下表：

参　　数	含　　义
"en"	指定文本的语言是英语
sampleStream	训练文本流
true	指定是否应该使用所示结束标记
null	一个缩略词词典
TrainingParameters.defaultParams()	指定使用默认训练参数

在下列代码中，创建了一个 OutputStream，并用它来保存模型到 modeFile 文件中，可以使这个模型在其他应用中重复利用：

```
OutputStream modelStream = new BufferedOutputStream(
    new FileOutputStream("modelFile"));
model.serialize(modelStream);
```

结果如下。为了节省空间，所有的迭代次数没有全部列出。默认中断索引到 5，并迭代 100 次。

```
Indexing events using cutoff of 5

    Computing event counts...  done. 93 events
    Indexing...  done.
Sorting and merging events... done. Reduced 93 events to 63.
Done indexing.
Incorporating indexed data for training...
done.
    Number of Event Tokens: 63
        Number of Outcomes: 2
        Number of Predicates: 21
...done.
Computing model parameters ...
Performing 100 iterations.
  1:  ... loglikelihood=-64.4626877920749      0.9032258064516129
  2:  ... loglikelihood=-31.11084296202819     0.9032258064516129
  3:  ... loglikelihood=-26.418795734248626    0.9032258064516129
  4:  ... loglikelihood=-24.327956749903198    0.9032258064516129
  5:  ... loglikelihood=-22.766489585258565    0.9032258064516129
  6:  ... loglikelihood=-21.46379347841989     0.9139784946236559
  7:  ... loglikelihood=-20.356036369911394    0.9139784946236559
  8:  ... loglikelihood=-19.406935608514992    0.9139784946236559
  9:  ... loglikelihood=-18.58725539754483     0.9139784946236559
 10:  ... loglikelihood=-17.873030559849326    0.9139784946236559
...
 99:  ... loglikelihood=-7.214933901940582     0.978494623655914
100:  ... loglikelihood=-7.183774954664058     0.978494623655914
```

3.6.1 使用训练好的模型

如下所示，我们来使用这个模型。这是基于在 3.5.1.1 节介绍的方法。

```java
try (InputStream is = new FileInputStream(
        new File(getModelDir(), "modelFile"))) {
    SentenceModel model = new SentenceModel(is);
    SentenceDetectorME detector = new
    SentenceDetectorME(model);
    String sentences[] = detector.sentDetect(paragraph);
    for (String sentence : sentences) {
        System.out.println(sentence);
    }
} catch (FileNotFoundException ex) {
    // Handle exception
```

```
} catch (IOException ex) {
    // Handle exception
}
```

结果如下：

```
When determining the end of sentences we need to consider several
factors.
Sentences may end with exclamation marks! Or possibly questions marks?
Within sentences we may find numbers like 3.14159,
abbreviations such as found in Mr.
Smith, and possibly ellipses either within a sentence …, or at the end of
a sentence…
```

这个模型并没有很好地处理最后一个句子，它反映了样本文本和模型使用的文本的不匹配。因此，使用相关的训练数据非常重要，否则，基于这个结果的下游任务将会遭殃。

3.6.2 使用 SentenceDetectorEvaluator 类评估模型

我们保留了一部分样本文件用于评估，因此我们能使用 SentenceDetecorEvaluator 类评估模型。我们更改 sentence.train 文件，提取最后 10 个句子放在另外一个文件 evalSample 中，并使用这个文件来评估模型。下面的示例中，我们基于文件内容，重用了 lineStream 和 sampleStream 变量创建一个 SentenceSample 对象流。

```
lineStream = new PlainTextByLineStream(
    new FileReader("evalSample"));
sampleStream = new SentenceSampleStream(lineStream);
```

使用之前创建的 SentenceDetectorME 类变量 detector 创建一个 SentenceDetectorEvaluator 类的实例。构造器的第二个参数是一个 SentenceDetectorEvaluationMonitor 对象（此处没有用到）。然后调用 evaluate 方法：

```
SentenceDetectorEvaluator sentenceDetectorEvaluator
    = new SentenceDetectorEvaluator(detector, null);
sentenceDetectorEvaluator.evaluate(sampleStream);
```

getFMeasure 方法将返回一个 FMeasure 类的实例，它提供了模型性能的度量：

```
System.out.println(sentenceDetectorEvaluator.getFMeasure());
```

结果如下。查准率是包含正确实例的比例，召回率反映了模型的敏感性。F 值是一个综合查准率和召回率的得分，从本质上反映了模型的工作性能。在分词和 SBD 任务上，最好保持查准率在 90% 以上：

```
Precision: 0.8181818181818182
Recall: 0.9
F-Measure: 0.8571428571428572
```

3.7 本章小结

我们讨论了一些文本断句问题的难点所在。问题包括句号用于数字、缩略词的场合。省略号的使用和嵌入的引号也都会产生问题。

Java 提供了一些判断句子结束的方法。可以使用正则表达式和 BreakIterator 类，这些方法对于简单句子非常有用，但是它们对那些稍微复杂的句子无能为力。

本章还介绍了各种 NLP API。有一些基于规则，有一些使用模型。我们也说明了模型怎样训练和评估。

在下一章，你将学到如何进行人物识别。

第 4 章 人物识别

寻找人和物的过程叫作命名实体识别（NER）。实体（如人和地名等）是那些有名字、区别于其他的类别，常见的实体类型包括：

- 人
- 位置
- 组织
- 钱
- 时间
- URL

在一个文档中找出名字、位置以及各类实体是重要且实用的 NLP 任务。它们在很多地方都有用到，比如创建简单的搜索，提问处理，解析目录，文本消歧，以及寻找文本的含义。例如，有时候用 NER 来判断属于一个简单类别的实体。通过分类，搜索过程就能分离这些项目类型。其他 NLP 任务也用到 NER，比如 POS 标注或交叉索引任务。

NER 处理涉及两种任务：

- 实体检测
- 实体分类

实体检测和寻找文本中的实体位置相关。一旦被定位，判断发现的实体所属类别尤为重要。完成这两个任务后，其结果能用于解决其他类似搜索和判断文本含义之类的任务。打个比方，从一部电影或者书籍的评论中识别名字来帮助找到可能感兴趣的其他电影或者书籍，提取到位置信息能够帮助提供附件服务的参考信息。

4.1 NER 难在何处

像很多 NLP 任务一样，NER 并不那么简单。尽管文本分词展示出了它的组成元素，但理解这些元素是什么依然很困难。语言的歧义性导致即使使用专有名词也不一定能理解其含义。例如 Penny 和 Faith，作为真实的姓名，但也可能表达财物度量和信仰。同样，如 Georgia 可以用来命名乡镇、行政区和人。

一些短语理解起来很有挑战性。"城市会议展览"中就包含有效的实体。当熟知实体所属领域时，实体列表就非常有用且易于实现 NER。

NER 在句子层面的应用很典型，一个短语很容易跨越一个句子导致一个实体识别错误。例如，下行句子中：

"Bob went south. Dakota went west."

如果忽视句子的边界，那么我们意外地发现南达科他州（美国中北部州）这一位置实体。特别的文档（比如 URL、邮箱地址）也很难从文中分离。如果再考虑实体形式的多样性，那么识别变得更为艰难。例如，电话号码中使用括号吗？号码中有没有使用破折号、句号或者一些其他的符号使其隔开？我们需要考虑国际电话号码吗？

这些因素带来了对成熟的 NER 技术的需求。

4.2 NER 的方法

有许多可供选择的 NER 方法，一些使用正则表达式，另一些基于预定义的字典。正则表达式有丰富的表达能力，能够分离实体。实体名字的字典能和文本中的词项通过对比进行匹配。

另一个常用的 NER 方法是使用训练模型来检测实体的存在，这些模型依赖于所要寻找的实体以及目标语言，适用于某一个领域的模型（如网页）可能不适用于另一个领域（如媒体新闻）。

训练模型时使用文本中已识别出实体的注释块。为了测试训练模型的好坏，可以使用以下几个方法。

- 查准率：数据中模型找到的匹配正确的实体数目与找到的所有实体数目的百分比。
- 召回率：找到的语料库中所定义的实体所占百分比。
- 性能度量：综合查准率和召回率的一种度量方法，定义为 F1 = 2 × 查准率 × 召回率 / (查准率 + 召回率)。

衡量模型好坏的时候将会用到这些测量方式。

NER 也常用于实体识别和实体分块。分块是判断文本中一些诸如名词、动词或者其他成分的分析过程。人们趋向于把句子划分成独立的部分，这些部分形成一个决定句子含义的结构。NER 处理将创建文本的跨越性实体，例如"Queen of England"。然而，在这些跨越性文本里有其他类似"England"的实体。

4.2.1 列表和正则表达式

命名实体识别（如专有名词）的一种方法是使用"标准"实体和正则表达式。标准实体列表由一系列乡镇、通用名、月份或者频繁引用的地理位置组成。常用地名表(gazetteer)，是一个包含配合地图使用的地理位置信息，能提供位置相关实体的列表。然而，维护这样的列表需要时间。它们也可能用于特定的语言和场景。改动这些列表很枯燥乏味。4.4.3.2 节将说明这个方法。

在识别实体的时候，正则表达式很有用。强大的句法为很多场合提供足够的灵活性来精确划分感兴趣的实体。然而，灵活性也可能导致难以理解和掌握。本章将论述几个正则表达式方法。

4.2.2 统计分类器

统计分类器决定一个单词是否是一个实体的开头、一个实体的延续，或者不是一个实体。示例文本中标记了分离的实体。分类器能针对不同问题领域用不同数据集来训练，这个方法的缺陷是需要有人注释示例文本，这需要花费时间，另外还和问题领域相关。

我们将演示执行 NER 的几个方法。首先，从解释正则表达式如何用于识别实体开始。

4.3 使用正则表达式进行 NER

正则表达式能用来识别文献中的实体。我们将调查两个常用的途径:

- 第一个方法使用 Java 支持的正则表达式,当实体相对简单而且具有统一形式时,此方法很有用。
- 第二个方法使用为特殊用途定制的正则表达式类。我们将使用 LingPipe 的 RegExChunker 类说明这一方法。

用正则表达式的方式可以利用前人已研究出的成果,有许多可用的预定义和测试过的表达式的来源,从 http://regexlib.com/Default.aspx 中能找到这样一个库。我们将使用这个库中的几个正则表达式作为例子。

为了测试这些方法的性能,大多数例子中会用到如下文本:

```
private static String regularExpressionText
    = "He left his email address (rgb@colorworks.com) and his "
    + "phone number,800-555-1234. We believe his current address "
    + "is 100 Washington Place, Seattle, CO 12345-1234. I "
    + "understand you can also call at 123-555-1234 between "
    + "8:00 AM and 4:30 most days. His URL is http://example.com "
    + "and he was born on February 25, 1954 or 2/25/1954.";
```

4.3.1 使用 Java 的正则表达式来寻找实体

为了解释这些表达式的使用,从几个简单的例子开始。这些例子以如下声明开头。这是为识别电话号码类型而设计的一个简单的表达式:

```
String phoneNumberRE = "\\d{3}-\\d{3}-\\d{4}";
```

使用如下代码来测试表达式。Pattern 类的 compile 方法接受一个正则表达式并把它编译成 Pattern 对象。然后它的 matcher 方法能执行目标文本,返回一个 Matcher 对象。这个对象允许我们重复识别相匹配的正则表达式:

```
Pattern pattern = Pattern.compile(phoneNumberRE);
Matcher matcher = pattern.matcher(regularExpressionText);
while (matcher.find()) {
    System.out.println(matcher.group() + " [" + matcher.start()
        + ":" + matcher.end() + "]");
}
```

当匹配成功时,find 方法将返回 true。它的 group 方法返回匹配这个表达式的文本。start 和 end 方法提供匹配文本在目标文本的位置。

完成后，将得到以下结果：

800-555-1234 [68:80]

123-555-1234 [196:208]

其他的许多正则表达式用法相近。这些已经在下表中列出。第三列是相应的正则表达式用在上面的代码中执行后的结果：

实体类型	正则表达式	结果
URL	\\b(https?|ftp|file|ldap)://[-A-Za-z0-9+&@#/%?=~_|!:,.;]*[-AZa-z0-9+&@#/%=~_|]	http://example.com [256:274]
邮政编码	[0-9]{5}(\\-?[0-9]{4})?	12345-1234[150:160]
邮箱	[a-zA-Z0-9'._%+-]+@(?:[a-zA-Z0-9-]+\\.)+[a-zA-Z]{2,4}	rgb@colorworks.com [27:45]
时间	(((0-1]?[0-9])|([2][0-3])):([0-5][0-9])(:([0-5]?[0-9]))?	8:00 [217:221] 4:30 [229:233]
日期	((0?[13578]|10|12)(-|\\/)(([1-9])|(0[1-9])|([12])([0-9]?)|(3[01]?))(-|\\/)((19)([2-9])(\\d{1})|(20)([01])(\\d{1})|([8901])(\\d{1}))|(0?[2469]|11)(-|\\/)(([1-9])|(0[1-9])|([12])([0-9]?)|(3[0]?))(-|\\/)((19)([2-9])(\\d{1})|(20)([01])(\\d{1})|([8901])(\\d{1})))	2/25/1954 [315:324]

我们可能还会用到许多其他的正则表达式。然而，这些例子阐述了基本方法。像日期正则表达式所示那样，一些正则表达式非常复杂。

正则表达式常遗漏一些实体或者把非实体误认为实体。例如，如果用下述表达式代替文本：

regularExpressionText =
    "(888)555-1111 888-SEL-HIGH 888-555-2222-J88-W3S";

执行代码将返回：

888-555-2222 [27:39]

它遗漏了开始的两个电话号码数字，而且把"区域编号"误认为电话号码。

我们也可以使用"|"操作符同时搜索两个或两个以上的正则表达式。在下面的表达式中，三个正则表达式通过这个操作符组合在一起，并依照上述表格中相应的条目声明：

```
Pattern pattern = Pattern.compile(phoneNumberRE + "|"
    + timeRE + "|" + emailRegEx);
```

当使用前面定义过的原始 regularExpressionText 文本执行时,我们得到了以下结果:

```
rgb@colorworks.com [27:45]
800-555-1234 [68:80]
123-555-1234 [196:208]
8:00 [217:221]
4:30 [229:233]
```

4.3.2 使用 LingPipe 的 RegExChunker 类

RegExChunker 类使用分块来寻找文本中的实体。该类使用一个正则表达式来代表一个实体。它的 chunk 方法返回一个与之前实例中所用实体使用方法一样的 Chunking 对象。

RegExChunker 类的构造器接受三个参数。

- String:一个正则表达式
- String:实体或者目录的类型
- double:分数值

我们将使用一个表示时间的正则表达式来解释这个类,如下所示,这个正则表达式与 4.3.1 节所使用的正则表达式相同。然后创建 Chunker 实例:

```
String timeRE =
    "(([0-1]?[0-9])|([2][0-3])):([0-5]?[0-9])(:([0-5]?[0-9]))?";
        Chunker chunker = new RegExChunker(timeRE,"time",1.0);
```

chunk 方法同 displayChunkset 方法一起使用,如下:

```
Chunking chunking = chunker.chunk(regularExpressionText);
Set<Chunk> chunkSet = chunking.chunkSet();
displayChunkSet(chunker, regularExpressionText);
```

displayChunkset 方法如下所示。chunkSet 方法返回一个关于 Chunk 实例的 Set 集合。可以使用各种方法来显示块中指定部分:

```
public void displayChunkSet(Chunker chunker, String text) {
    Chunking chunking = chunker.chunk(text);
    Set<Chunk> set = chunking.chunkSet();
    for (Chunk chunk : set) {
        System.out.println("Type: " + chunk.type() + " Entity: ["
            + text.substring(chunk.start(), chunk.end())
```

```
            + "] Score: " + chunk.score());
    }
}
```

结果如下：

Type: time Entity: [8:00] Score: 1.0
Type: time Entity: [4:30] Score: 1.0+95

此外，可以声明一个简单的类来封装其他场合重用的正则表达式，接下来，声明 TimeRegexChunker 类，它支持时间实体的识别：

```
public class TimeRegexChunker extends RegExChunker {
    private final static String TIME_RE =
        "(([0-1]?[0-9])|([2][0-3])):([0-5]?[0-9])(:([0-5]?[0-9]))?";
    private final static String CHUNK_TYPE = "time";
    private final static double CHUNK_SCORE = 1.0;

    public TimeRegexChunker() {
        super(TIME_RE,CHUNK_TYPE,CHUNK_SCORE);
    }
}
```

为使用这个类，用以下声明代替这一节中 chunker 的初始声明：

```
Chunker chunker = new TimeRegexChunker();
```

结果与之前相同。

4.4 使用 NLP API

我们将使用 OpenNLP、Stanford API 和 LingPipe 来展示 NER 处理。这三个都提供了良好的文本实体识别方法。下述声明将用作示例文本说明这些 API：

```
String sentences[] = {"Joe was the last person to see Fred. ",
    "He saw him in Boston at McKenzie's pub at 3:00 where he "
    + " paid $2.45 for an ale. ",
    "Joe wanted to go to Vermont for the day to visit a cousin who "
    + "works at IBM, but Sally and he had to look for Fred"};
```

4.4.1 使用 OpenNLP 进行 NER

本节将论述如何基于 OpenNLP API 使用 TokenNameFinderModel 类进行 NLP，另外，还将演示如何确定实体识别的正确率。

通常做法是将文本转换为一系列独立句子，使用一个合适的模型来创建 TokenName-

FinderModel 类的一个实例,然后使用 find 方法来识别文本中的实体。

以下例子论述了 TokenNameFinderModel 类的使用。实例先针对一个简单句再使用复杂句。句子如下:

```
String sentence = "He was the last person to see Fred.";
```

我们将分别使用 en-token.bin 文件和 en-ner-person.bin 文件中找到的模型作为分词器以及名字查找器模型。用如下 try-with-resources 语句块打开用于这些文件的 InputStream 对象:

```
try (InputStream tokenStream = new FileInputStream(
        new File(getModelDir(), "en-token.bin"));
        InputStream modelStream = new FileInputStream(
            new File(getModelDir(), "en-ner-person.bin"));) {
    ...

} catch (Exception ex) {
    // Handle exceptions
}
```

在 try 块中创建 TokenizerModel 和 Tokenizer 对象:

```
TokenizerModel tokenModel = new TokenizerModel(tokenStream);
Tokenizer tokenizer = new TokenizerME(tokenModel);
```

接下来,使用个人模型创建 NameFinderME 类的一个实例:

```
TokenNameFinderModel entityModel =
    new TokenNameFinderModel(modelStream);
NameFinderME nameFinder = new NameFinderME(entityModel);
```

现在可以使用 tokenize 方法进行文本分词,并使用 find 方法来识别文本中的人。find 方法将使用分词所得 String 数组作为输入,并返回如下的 Span 对象数组:

```
String tokens[] = tokenizer.tokenize(sentence);
Span nameSpans[] = nameFinder.find(tokens);
```

你可能还记得在第 3 章中讨论过的 Span 类,这个类包含已找到实体的位置信息。实际的字符串实体仍然存在于 tokens 数组中。

如下语句显示句子中找到的人。其位置信息和人分别显示在不同的行中:

```
for (int i = 0; i < nameSpans.length; i++) {
    System.out.println("Span: " + nameSpans[i].toString());
    System.out.println("Entity: "
        + tokens[nameSpans[i].getStart()]);
}
```

结果如下:

Span: [7..9) person
Entity: Fred

我们经常和复杂句打交道。为了演示复杂句,使用先前定义的 sentences 字符串数组,并用以下语句代替前面的声明。逐句调用 tokenize 方法,然后以同样的方式列出实体信息:

```
for (String sentence : sentences) {
    String tokens[] = tokenizer.tokenize(sentence);
    Span nameSpans[] = nameFinder.find(tokens);
    for (int i = 0; i < nameSpans.length; i++) {
        System.out.println("Span: " + nameSpans[i].toString());
        System.out.println("Entity: "
            + tokens[nameSpans[i].getStart()]);
    }
    System.out.println();
}
```

结果如下。由于第二个句子中不包含人,因此识别到的两个人中间会多出一行空白:

Span: [0..1) person
Entity: Joe
Span: [7..9) person
Entity: Fred

Span: [0..1) person
Entity: Joe
Span: [19..20) person
Entity: Sally
Span: [26..27) person
Entity: Fred

4.4.1.1 实体识别的准确度

TokenNameFinderModel 识别文本中实体的同时还计算了实体识别的可能性。我们能使用如下行代码中的 probs 方法获取这个信息。这个方法返回一个对应于 nameSpans 数组元素的 double 数组:

```
double[] spanProbs = nameFinder.probs(nameSpans);
```

使用 find 方法后直接把这行语句添加到之前例子中,然后在嵌套的 for 语句的最后添加如下一行：

```
System.out.println("Probability: " + spanProbs[i]);
```

执行后将得到以下结果。这个概率反映了指定实体的置信度。对第一个实体来说，模型有 80.529% 的把握确信"Joe"是一个人：

```
Span: [0..1) person
Entity: Joe
Probability: 0.8052914774025202
Span: [7..9) person
Entity: Fred
Probability: 0.9042160889302772

Span: [0..1) person
Entity: Joe
Probability: 0.9620970782763985
Span: [19..20) person
Entity: Sally
Probability: 0.964568603518126
Span: [26..27) person
Entity: Fred
Probability: 0.990383039618594
```

4.4.1.2 使用其他实体类型

OpenNLP 支持以下列表中列出的不同的库。这些模型能从 http://opennlp.sourceforge.net/models-1.5/ 下载。前缀 en 说明默认语言为英语，ner 表示用于 NER 的模型。

英语查找模型	文件名
Location name finder model	en-ner-location.bin
Money name finder model	en-ner-money.bin
Organization name finder model	en-ner-organization.bin
Percentage name finder model	en-ner-percentage.bin
Person name finder model	en-ner-person.bin
Time name finder model	en-ner-time.bin

如果要为使用一个不同的模型文件而修改语句，可以参考以下例句：

```
InputStream modelStream = new FileInputStream(
    new File(getModelDir(), "en-ner-time.bin"));) {
```

当使用 en-ner-money.bin 模型时，先前代码行中的 tokens 数组索引必须增加 1，否则返回值都是美元符号。

下表展示了各类模型的结果。

模　型	输　　出
en-ner-location.bin	Span: [4..5) location Entity: Boston Probability: 0.8656908776583051 Span: [5..6) location Entity: Vermont Probability: 0.9732488014011262
en-ner-money.bin	Span: [14..16) money Entity: 2.45 Probability: 0.7200919701507937
en-ner-organization.bin	Span: [16..17) organization Entity: IBM Probability: 0.9256970736336729
en-ner-time.bin	The model was not able to detect time in this text sequence

示例文本不能查找时间实体，说明这个模型不足以确定文本中找到的实体是时间。

4.4.1.3　处理多种实体类型

我们也能同时处理多种实体类型。这涉及在循环中创建一个基于各模型的 NameFinderME 类的实例，逐句应用模型并记录找到的实体。

我们将用下面的实例解释这个处理过程。它需要重写之前的 try 块以便在块中创建 InputStream 实例，如下：

```
try {
    InputStream tokenStream = new FileInputStream(
        new File(getModelDir(), "en-token.bin"));
    TokenizerModel tokenModel = new TokenizerModel(tokenStream);
    Tokenizer tokenizer = new TokenizerME(tokenModel);
    ...
} catch (Exception ex) {
    // Handle exceptions
}
```

try 块中定义了一个记录模型文件名的字符串数组。如下所示，使用人、地理位置以及组织这三个模型：

```
String modelNames[] = {"en-ner-person.bin",
    "en-ner-location.bin", "en-ner-organization.bin"};
```

创建一个 ArrayList 实例来保存发现的实体：

```
ArrayList<String> list = new ArrayList();
```

使用 for-each 语句一次加载一个模型，然后创建一个 NameFinderME 类的实例：

```
for(String name : modelNames) {
    TokenNameFinderModel entityModel = new TokenNameFinderModel(
        new FileInputStream(new File(getModelDir(), name)));
    NameFinderME nameFinder = new NameFinderME(entityModel);
    ...
}
```

先前我们并没有识别所找到的实体具体来自哪个句子，但这并不难实现，只需要使用一个简单的 for 语句代替 for-each 语句来追踪句子的索引。如下例所示，在前例基础上使用整型变量 index 来保留句子，否则，代码运行结果和之前一样：

```
for (int index = 0; index < sentences.length; index++) {
    String tokens[] = tokenizer.tokenize(sentences[index]);
    Span nameSpans[] = nameFinder.find(tokens);
    for(Span span : nameSpans) {
        list.add("Sentence: " + index
            + " Span: " + span.toString() + " Entity: "
            + tokens[span.getStart()]);
    }
}
```

随后列出找到的实体：

```
for(String element : list) {
    System.out.println(element);
}
```

结果如下：

```
Sentence: 0 Span: [0..1) person Entity: Joe
Sentence: 0 Span: [7..9) person Entity: Fred
Sentence: 2 Span: [0..1) person Entity: Joe
Sentence: 2 Span: [19..20) person Entity: Sally
Sentence: 2 Span: [26..27) person Entity: Fred
Sentence: 1 Span: [4..5) location Entity: Boston
Sentence: 2 Span: [5..6) location Entity: Vermont
Sentence: 2 Span: [16..17) organization Entity: IBM
```

4.4.2 使用 Stanford API 进行 NER

我们将论述用于 NER 的 CRFClassifier 类。这个类可以实现马尔可夫链条件随机场（CRF）序列模型。

为了解释 CRFClassifier 类的使用，我们先声明分类器文件字符串，如下所示：

```
String model = getModelDir() +
    "\\english.conll.4class.distsim.crf.ser.gz";
```

然后使用模型创建分类器：

```
CRFClassifier<CoreLabel> classifier =
    CRFClassifier.getClassifierNoExceptions(model);
```

classify 方法输入值为一个表示待处理文本的字符串。因此为使用 sentences 文本，我们需要把它转换为一个简单的字符串：

```
String sentence = "";
for (String element : sentences) {
    sentence += element;
}
```

然后对文本应用 classify 方法：

```
List<List<CoreLabel>> entityList = classifier.classify(sentence);
```

返回 CoreLabel 对象众多 List 实例中的一个。返回对象是一个包含另一个列表的列表，被包含的列表是 CoreLabel 对象的一个 List 实例，CoreLabel 类表示一个带有附加信息的词。"内部"列表包含这些词的列表。在下述代码行中的 for-each 语句外部，引用变量 internalList 表示文本中的一个句子。在每个 for-each 语句内展示了内部列表中的每个词。word 方法返回单词，同时 get 方法返回单词的类型。然后列出单词及其类型：

```
for (List<CoreLabel> internalList: entityList) {
    for (CoreLabel coreLabel : internalList) {
        String word = coreLabel.word();
        String category = coreLabel.get(
            CoreAnnotations.AnswerAnnotation.class);
        System.out.println(word + ":" + category);
    }
}
```

部分结果如下。由于所有单词都需要显示，因此有所删减，O 代表"其他"类型：

```
Joe:PERSON
was:O
the:O
```

```
last:O
person:O
to:O
see:O
Fred:PERSON
.:O
He:O
...
look:O
for:O
Fred:PERSON
```

为了滤除不相关的单词，用以下语句代替 println 语句，这会剔除其他类型，只保留你想要的类型：

```
if (!"O".equals(category)) {
    System.out.println(word + ":" + category);
}
```

现在结果简单多了：

```
Joe:PERSON
Fred:PERSON
Boston:LOCATION
McKenzie:PERSON
Joe:PERSON
Vermont:LOCATION
IBM:ORGANIZATION
Sally:PERSON
Fred:PERSON
```

4.4.3 使用 LingPipe 进行 NER

本章前面的 4.3 节论述了 LingPipe 中正则表达式的使用。这里，我们将论述命名实体模型和 ExactDictionaryChunker 类如何用于 NER 分析。

4.4.3.1 使用 LingPipe 的命名实体模型

LingPipe 有一些可以和 chunking 一起使用的命名实体模型。这些文件由一个序列化的从一个文件中读出的对象组成，然后应用于文本。它们实现 Chunker 接口，chunking 过程生成一系列能对感兴趣的实体进行识别的 chunking 对象。

下表列出了一组 NER 模型，这些模型能从 http://alias-i.com/lingpipe/web/models.html 下载：

类　　型	语料库	文　　件
English News	MUC-6	ne-en-news-muc6.AbstractCharLmRescoringChunker
English Genes	GeneTag	ne-en-bio-genetag.HmmChunker
English Genomics	GENIA	ne-en-bio-genia.TokenShapeChunker

我们将使用文件 ne-en-news-muc6.AbstractCharLmRescoringChunker 中的模型说明这个类的使用方法。

如下所示，首先，我们以 try-catch 块处理异常。打开这个文件并使用 AbstractExternalizable 类的静态 readObject 方法来创建一个 Chunker 类的实例。该方法能读取序列化模型：

```
try {
    File modelFile = new File(getModelDir(),
        "ne-en-news-muc6.AbstractCharLmRescoringChunker");
    Chunker chunker = (Chunker)
        AbstractExternalizable.readObject(modelFile);
    ...
} catch (IOException | ClassNotFoundException ex) {
    // Handle exception
}
```

Chunker 和 Chunking 接口提供了处理文本的语块集合的方法。它的 chunk 方法返回一个实现 Chunking 实例的对象。如下所示为文本里每个句子中找到的语块：

```
for (int i = 0; i < sentences.length; ++i) {
    Chunking chunking = chunker.chunk(sentences[i]);
    System.out.println("Chunking=" + chunking);
}
```

结果如下：

```
Chunking=Joe was the last person to see Fred.  : [0-3:PERSON@-Infinity,
31-35:ORGANIZATION@-Infinity]
Chunking=He saw him in Boston at McKenzie's pub at 3:00 where he paid
$2.45 for an ale.  : [14-20:LOCATION@-Infinity, 24-32:PERSON@-Infinity]
Chunking=Joe wanted to go to Vermont for the day to visit a cousin who
works at IBM, but Sally and he had to look for Fred : [0-3:PERSON@-
Infinity, 20-27:ORGANIZATION@-Infinity, 71-74:ORGANIZATION@-Infinity,
109-113:ORGANIZATION@-Infinity]
```

另外，我们能使用 Chunk 类中的方法从所示信息中提取特定的信息片段。用如下 for-each 语句代替前面的语句。本章前面 4.3.2 节已经论述了这个 displayChunkSet 方法：

```
for (String sentence : sentences) {
    displayChunkSet(chunker, sentence);
}
```

结果如下。然而结果类型并不总是和正确的实体类型相匹配。

```
Type: PERSON Entity: [Joe] Score: -Infinity
Type: ORGANIZATION Entity: [Fred] Score: -Infinity
Type: LOCATION Entity: [Boston] Score: -Infinity
Type: PERSON Entity: [McKenzie] Score: -Infinity
Type: PERSON Entity: [Joe] Score: -Infinity
Type: ORGANIZATION Entity: [Vermont] Score: -Infinity
Type: ORGANIZATION Entity: [IBM] Score: -Infinity
Type: ORGANIZATION Entity: [Fred] Score: -Infinity
```

4.4.3.2　使用 ExactDictionaryChunker 类

ExactDictionaryChunker 类提供了一个简单方法来创建实体及其类型的字典。这个字典能用于文本中寻找实体及其类型。它使用一个 MapDictionary 对象来存储实体，然后根据这个字典使用 ExactDictionaryChunker 类来提取语块。

AbstractDictionary 接口支持实体、类别和分数的基本操作。这里的分数用作匹配处理。MapDictionary 和 TrieDictionary 类是 AbstractDictionary 的两个接口，TrieDictionary 类使用一个字符字典树结构存储信息。这种方法占用更少内存，我们将以 MapDictionary 类为例。

为了解释这个方法，我们以一个 MapDictionary 类的声明开头：

```
private MapDictionary<String> dictionary;
```

这个字典包含了我们想找到的实体。使用 initializeDictionary 方法对模型进行初始化。这里使用的 DictionaryEntry 构造器接受三类参数：

- String：实体名
- String：实体类型
- Double：代表实体的分数

确定是否匹配时会用到分数，声明一部分实体，然后添加到字典。

```
private static void initializeDictionary() {
    dictionary = new MapDictionary<String>();
    dictionary.addEntry(
        new DictionaryEntry<String>("Joe","PERSON",1.0));
    dictionary.addEntry(
        new DictionaryEntry<String>("Fred","PERSON",1.0));
    dictionary.addEntry(
        new DictionaryEntry<String>("Boston","PLACE",1.0));
    dictionary.addEntry(
        new DictionaryEntry<String>("pub","PLACE",1.0));
    dictionary.addEntry(
        new DictionaryEntry<String>("Vermont","PLACE",1.0));
    dictionary.addEntry(
        new DictionaryEntry<String>("IBM","ORGANIZATION",1.0));
    dictionary.addEntry(
        new DictionaryEntry<String>("Sally","PERSON",1.0));
}
```

ExactDictionaryChunker 实例会使用这个字典。ExactDictionaryChunker 类参数说明如下：

- Dictionary<String>：一个包含实体的字典
- TokenizerFactory：chunker 使用的一个分词器
- boolean：若为 true，chunker 返回所有匹配的值
- boolean：若为 true，匹配内容区分大小写

匹配内容允许重叠。例如，短语"第一国家银行"（The First National Bank），银行实体能独立使用，也能和短语中的其他成分配合使用，第三个参数决定是否返回所有的匹配值。

下列代码中，字典进行了初始化。然后我们使用印欧语系的分词器创建一个 ExactDictionaryChunker 类的实例，这里返回所有的匹配值，并忽略词项的大小写：

```
initializeDictionary();
ExactDictionaryChunker dictionaryChunker
    = new ExactDictionaryChunker(dictionary,
        IndoEuropeanTokenizerFactory.INSTANCE, true, false);
```

逐句使用 dictionaryChunker 对象。如以下代码所示，我们将使用 4.3.2 节所提到的 displayChunkSet 方法：

```
for (String sentence : sentences) {
    System.out.println("\nTEXT=" + sentence);
    displayChunkSet(dictionaryChunker, sentence);
}
```

执行完，将得到以下结果：

```
TEXT=Joe was the last person to see Fred.
Type: PERSON Entity: [Joe] Score: 1.0
Type: PERSON Entity: [Fred] Score: 1.0

TEXT=He saw him in Boston at McKenzie's pub at 3:00 where he paid $2.45
for an ale.
Type: PLACE Entity: [Boston] Score: 1.0
Type: PLACE Entity: [pub] Score: 1.0
TEXT=Joe wanted to go to Vermont for the day to visit a cousin who works
at IBM, but Sally and he had to look for Fred
Type: PERSON Entity: [Joe] Score: 1.0
Type: PLACE Entity: [Vermont] Score: 1.0
Type: ORGANIZATION Entity: [IBM] Score: 1.0
Type: PERSON Entity: [Sally] Score: 1.0
Type: PERSON Entity: [Fred] Score: 1.0
```

效果很不错，但需要很大的精力创建包含大量词汇的字典。

4.5 训练模型

我们将使用 OpenNLP 来论述如何训练一个模型。训练文件必须满足以下要求：

- 包含区分实体边界的标注
- 每行一个句子

使用文件名为 en-ner-person.train 的模型文件：

```
<START:person> Joe <END> was the last person to see <START:person>
Fred <END>.
He saw him in Boston at McKenzie's pub at 3:00 where he paid $2.45 for
an ale.
<START:person> Joe <END> wanted to go to Vermont for the day to visit
a cousin who works at IBM, but <START:person> Sally <END> and he had
to look for <START:person> Fred <END>.
```

示例中的几个方法都能抛出异常信息。上面的这些语句将放入如下的 try-with-resource 块中，同时创建模型的输出流：

```
try (OutputStream modelOutputStream = new BufferedOutputStream(
        new FileOutputStream(new File("modelFile")));) {
    ...
} catch (IOException ex) {
```

```
    // Handle exception
}
```

在 try-with-resource 块内，使用 PlainTextByLineStream 类创建 OutputStream<String> 对象。这个类的构造器调用 FileInputStream 实例，并把每一行作为一个 String 对象返回。en-ner-person.train 文件用作输入文件，像这里展示的一样，UTF-8 是所使用的编码序列：

```
ObjectStream<String> lineStream = new PlainTextByLineStream(
    new FileInputStream("en-ner-person.train"), "UTF-8");
```

lineStream 对象包含用标注的文本里描写的实体的数据流。这些需要转换为用于训练模型的 NameSample 对象。这个转换由如下显示的 NameSampleDataStream 类完成。一个 NameSample 对象记录了文中已发现实体的名字：

```
ObjectStream<NameSample> sampleStream =
    new NameSampleDataStream(lineStream);
```

通过如下程序执行 train 方法：

```
TokenNameFinderModel model = NameFinderME.train(
    "en", "person",  sampleStream,
    Collections.<String, Object>emptyMap(), 100, 5);
```

下表详细显示了这个方法的参数：

参数	含义
"en"	语言
"person"	实体类型
sampleStream	采样数据
null	资源
100	迭代的次数
5	截止条件

然后模型序列化到一个输出文件：

```
model.serialize(modelOutputStream);
```

结果如下。结果有所省略以节省空间。关于模型创建的基本信息如下：

```
Indexing events using cutoff of 5

   Computing event counts...  done. 53 events
   Indexing...  done.
```

```
Sorting and merging events... done. Reduced 53 events to 46.
Done indexing.
Incorporating indexed data for training...
done.
    Number of Event Tokens: 46
        Number of Outcomes: 2
      Number of Predicates: 34
...done.
Computing model parameters ...
Performing 100 iterations.
  1:  ... loglikelihood=-36.73680056967707    0.05660377358490566
  2:  ... loglikelihood=-17.499660626361216   0.9433962264150944
  3:  ... loglikelihood=-13.216835449617108   0.9433962264150944
  4:  ... loglikelihood=-11.461783667999262   0.9433962264150944
  5:  ... loglikelihood=-10.380239416084963   0.9433962264150944
  6:  ... loglikelihood=-9.570622475692486    0.9433962264150944
  7:  ... loglikelihood=-8.919945779143012    0.9433962264150944
...
 99:  ... loglikelihood=-3.513810438211968    0.9622641509433962
100:  ... loglikelihood=-3.507213816708068    0.9622641509433962
```

模型评估

模型可以使用 TokenNameFinderEvaluator 类进行评估。评估使用标注的示例文本。对于这个简单的示例，先创建包含如下文本的 en-ner-person.eval 文件：

```
<START:person> Bill <END> went to the farm to see <START:person> Sally
<END>.
Unable to find <START:person> Sally <END> he went to town.
There he saw <START:person> Fred <END> who had seen <START:person>
Sally <END> at the book store with <START:person> Mary <END>.
```

接下来的代码用来实施评估。先前的模型用作 TokenNameFinderEvaluator 构造器的参数。基于评估文件创建 NameSampleDataStream 实例。用 TokenNameFinderEvaluator 类的 evaluate 方法进行评估：

```
TokenNameFinderEvaluator evaluator =
    new TokenNameFinderEvaluator(new NameFinderME(model));
lineStream = new PlainTextByLineStream(
    new FileInputStream("en-ner-person.eval"), "UTF-8");
sampleStream = new NameSampleDataStream(lineStream);
evaluator.evaluate(sampleStream);
```

为了判断模型在测试数据中的表现，执行 getFMeasure 方法，结果如下：

```
FMeasure result = evaluator.getFMeasure();
System.out.println(result.toString());
```

如下的结果显示了查准率、召回率和 F 值，表明找到的实体有 50% 与测试数据吻合。召回率是找到的语料库中所定义的实体所占百分比。性能度量综合了查准率和召回率，定义为 F1 = 2× 查准率 × 召回率 /（查准率 + 召回率）

```
Precision: 0.5
Recall: 0.25
F-Measure: 0.3333333333333333
```

为创建一个更好的模型，数据和测试集应尽量大。注意这里只论述了训练和评估 POS 模型的基本方法。

4.6 本章小结

NER 涉及实体检测和实体分类。一般的类别包括名字、地理位置和事物。许多应用都用它为搜索引擎提供技术支持，解决参考文献以及判断文本的含义。这个过程经常用于下游任务。

我们探究了几个实现 NER 的技术。正则表达式是一个同时支持 Java 核心类和 NLP API 的方法。这个技术在很多应用中都有用，目前有很多可用的正则表达式库。

基于字典的方法在一些应用中也很有效。但有时需要大量精力完善字典。我们使用 LingPipe 的 MapDictionary 类说明了这个方法。

训练模型也能用来进行 NER。我们测试了一些模型，并论述了如何使用 OpenNLP NameFinderME 来训练模型。这个过程和之前训练过程相似。

下一章，我们将学习如何判断词性，如名词、形容词和介词。

CHAPTER 5

第 5 章

词性判断

在此之前，我们识别文本中的实体（如人、地点和事物）。在本章中，我们将研究词性判断的过程，包括在英语中识别语法元素，如名词和动词。我们将发现该单词的上下文是确定其类型的一个重要方面。

我们将研究标注过程，其本质上分配一个标注给词性，这个过程是词性判断的核心。我们将简要讨论标注的重要性，然后研究词性判断困难的各种影响因素。本章将使用各种 NLP 的 API 进行词性标注，还将说明如何训练一个针对特殊文本的模型。

5.1 词性标注

标注是对一个词汇或一段文字进行描述的过程。这个描述被称为一个标注。词性标注是分配 POS 标注给一个词项的过程。这些标注通常就是名词、动词和形容词等。

例如，思考下面的句子：

"The cow jumped over the moon."

对这样的例子，我们将演示使用 OpenNLP 标注词性的结果。讨论如何使用 OpenNLP 词性标注工具。如果我们使用标注前面的例子中的句子，我们会得到以下结果。请注意，单词后跟一个斜杠，然后跟着它们的词性标注。这些标注将简要说明：

```
The/DT cow/NN jumped/VBD over/IN the/DT moon./NN
```

根据上下文，单词会被分配多个标注。例如，单词"saw"可能是一个名词或动词。一个词可以被分类成不同的类别或信息，例如它的位置、相近的词，或相似性信息，这些信息可以确定其合适类别的概率。例如，如果一个词前面有一个定冠词之后又紧跟一个名词，那么这个词标注应为形容词。

一般的标注过程分为文本分词，判断可能的标注，解决不明确的标注，有一些算法可用于执行词性标识（标注）。一般有两种常用方法：

- 基于规则：基于规则标注器使用一组规则和词汇及其标注的字典。规则是在一个词汇具有多个标注时使用。规则通常使用词汇的上下文来选择一个标注。
- 随机概率：随机概率标注器或是基于马尔可夫模型或是线索模型的，它们根据基于决策树或最大熵原理。马尔可夫模型是有限状态机，其中每个状态有两种可能性分布，其目标是为一个句子找到标注的最优序列。隐马尔可夫模型（HMM）也可使用。在这些模型中，状态转换信息是不可见的。

最大熵标注器使用统计信息来确定词汇的词性，经常使用语料库来训练模型。语料库是被标注好词性的词汇的集合。语料库有多种语言。这些语料库花费了很大的力气来开发。经常使用的语料库包括 Penn Treebank（http://www.cis.upenn.edu/~treebank/）或 Brown 语料库（http://www.essex.ac.uk/linguistics/external/clmt/w3c/corpus_ling/content/corpora/list/private/brown/brown.html）。

这有一个来自从 Penn Treebank 语料库的例子演示了词性是如何标注的，如下：

```
Well/UH what/WP do/VBP you/PRP think/VB about/IN
the/DT idea/NN of/IN ,/, uh/UH ,/, kids/NNS having/VBG
to/TO do/VB public/JJ service/NN work/NN for/IN a/DT
year/NN ?/.
```

传统上英文有 9 个词性，包括：名词、动词、冠词、形容词、介词、代词、副词、连词和感叹词。然而，一个更完整的分析通常需要额外的类别和子类别。已经出现了多达 150 种不同词性标识。在某些情况下，有必要创建新的标注。如下是一个简短的标注列表。

这些都是本章中经常使用的标注：

标注	含义
NN	单数名词或集合名词
DT	限定词
VB	动词，基本形式
VBD	动词，过去式
VBZ	动词，第三人称单数现在时态
IN	介词或从属连词
NNP	单数专有名词
TO	to
JJ	形容词

更全面的内容如下表所示。此列表是来自 https://www.ling.upenn.edu/courses/Fall_2003/ling001/penn_treebank_pos.html。宾夕法尼亚大学（Penn）Treebank 标注集可以参考下面的网址 http://www.comp.leeds.ac.uk/ccalas/tagsets/upenn.html。一组标注被称为 tag set。

标注	说明	标注	说明
CC	并列连词	PRP$	物主代词
CD	基数	RB	副词
DT	限定词	RBR	副词，比较级
EX	存在句	RBS	副词，最高级
FW	外来词	RP	助词
IN	介词或从属连词	SYM	符号
JJ	形容词	TO	to
JJR	形容词，比较级	UH	感叹词
JJS	形容词，最高级	VB	动词，基本形式
LS	列表项标注	VBD	动词，过去式
MD	情态动词	VBG	动名词或现在分词
NN	单数名词或集合名词	VBN	动词，过去分词
NNS	复数名词	VBP	动词，非第三人称单数现在时态
NNP	单数专有名词	VBZ	动词，第三人称单数现在时态
NNPS	复数专有名词	WDT	wh- 限定词（wh 代表疑问词 who whose where when what）
PDT	前置限定词	WP	wh- 代名词
POS	所有格结束	WP$	所有格 wh- 代名词
PRP	人称代词	WRB	wh- 副词

人工语料库的开发是一项劳动密集的工作。然而，一些统计技术已经被开发用来创建语料库。一些语料库已经可以使用。其中首屈一指的是 Brown 语料库（http://clu.uni.no/icame/manuals/BROWN/INDEX.HTM）。比较新的语料库包括英国国家语料库（http://

www.natcorp.ox.ac.uk/corpus/index.xml），其已超过 100 万个单词，还有美国国家语料库（http://www.anc.org/）。语料库的列表可以在网址 http://en.wikipedia.org/wiki/List_of_text_corpora 找到。

5.1.1 词性标注器的重要性

正确标注的句子可以提高下游加工任务的质量。如果我们知道"sue"是一个动词而不是名词，那么这可以帮助确定词项之间的正确关系。确定在词性标注、短语、子句和它们之间的任何关系称为解析。与分词过程相反，分词只关心识别"word"元素而不关心它们的含义。

词性标注可用于许多下游流程，如问题的分析和文本的情感分析，一些社会媒体网站经常对其用户间交流的情感感兴趣，全文索引会频繁地使用的词性标注数据，语音处理可以使用标注来帮助决定如何发音。

5.1.2 词性标注难在何处

每一种语言都有很多方面特性可以使词性标注变得困难。大多数英语单词都会有两个或多个与之相关的标注。一本词典不足以确定某一个单词的词性。例如，像"bill"与"force"的单词其含义取决于它们的上下文。下面的句子演示了这两个单词是如果在同一个句子出现并分别作为名词和动词的。

"Bill used the force to force the manger to tear the bill in two."

使用 OpenNLP 来标注这句话得到下面的结果：

```
Bill/NNP used/VBD the/DT force/NN to/TO force/VB the/DT manger/NN to/TO
tear/VB the/DT bill/NN in/IN two./PRP$
```

社交媒介（如推特）使用的短信文（textese）是不同形式的文本的组合，包括缩写、标注、表情符号和俚语，使其更难以给句子进行标注。例如，下面的信息很难标注：

"AFAIK she H8 cth! BTW had a GR8 tym at the party BBIAM."

它相当于：

"As far as I know, she hates cleaning the house! By the way, had a great time at the party. Be back in a minute."

使用 OpenNLP 标注，我们将得到下面的结果：

AFAIK/NNS she/PRP H8/CD cth!/.
BTW/NNP had/VBD a/DT GR8/CD tym/NN at/IN the/DT party/NN BBIAM./.

在本章的后面 5.2.2.2 节，我们将示范如何使用 LingPipe 处理缩略语。下表中给出了缩略语简表：

短语	短信文	短语	短信文
As far as I know	AFAIK	By the way	BTW
Away from keyboard	AFK	You're on your own	YOYO
Thanks	THNX or THX	As soon as possible	ASAP
Today	2day	What do you mean by that	WDYMBT
Before	B4	Be back in a minute	BBIAM
See you	C U	Can't	CNT
Haha	hh	Later	l8R
Laughing out loud	LOL	On the other hand	OTOH
Rolling on the floor laughing	ROFL or ROTFL	I don't know	IDK
Great	GR8	Cleaning the house	CTH
At the moment	ATM	In my humble opinion	IMHO

有多个缩略语列表，一个比较全面的列表可在下面的网址中找到：http://www.ukrainecalling.com/textspeak.aspx。

分词是词性标注过程的重要一步。如果词项拆分不正确，则我们会得到错误的结果。此外，还有几个其他潜在的问题，包括：

- 如果我们使用小写，那么诸如"sam"会使得人和"系统管理奖"混淆（www.sam.gov）
- 我们必须考虑到缩略例如"can't"，并认识到会有不同的字符会用到单引号
- 例如短语"vice versa"可以视为一个单元，这是一个英格兰乐队名称、一部小说的标题、一本杂志的标题
- 我们不能忽视使用连字符的单词例如"first-cut"和"prime-cut"，它们的含义不同于其单个的含义

- 一些单词带有嵌入式数字如 iPhone 5s
- 特殊字符序列（如 URL 或电子邮件的地址）也需要处理

有些单词是嵌入在引号或括号中，会使它们的含义令人困惑。请看下面的示例：

"Whether 'Blue' was correct or not (it's not) is debatable."

"Blue"可能指的是蓝颜色或者可以想象成一个人的绰号。标注这句话的结果如下面所示：

```
Whether/IN "Blue"/NNP was/VBD correct/JJ or/CC not/RB (it's/JJ not)/NN
is/VBZ debatable/VBG
```

5.2　使用 NLP API

我们将使用 OpenNLP 库、Stanford API 库和 LingPipe 库来演示词性标注。每个示例将使用下面的句子。这是儒勒·凡尔纳的小说《海底两万里》第 5 章中的第一句话：

```
private String[] sentence = {"The", "voyage", "of", "the",
    "Abraham", "Lincoln", "was", "for", "a", "long", "time", "marked",
    "by", "no", "special", "incident."};
```

要处理的文本不可能总是以这种方式来定义。有时候句子会被定义为单个字符串：

```
String theSentence = "The voyage of the Abraham Lincoln was for a "
    + "long time marked by no special incident.";
```

我们可能需要将单个字符串转换为字符串数组。有很多方法可以把这个字符串转换为单词的数组。下面是使用 tokenizeSentence 方法执行这个操作：

```
public String[] tokenizeSentence(String sentence) {
    String words[] = sentence.split("S+");
    return words;
}
```

下面的代码将演示如何使用此方法：

```
String words[] = tokenizeSentence(theSentence);
for(String word : words) {
    System.out.print(word + " ");
}
System.out.println();
```

结果如下：

```
The voyage of the Abraham Lincoln was for a long time marked by no
special incident.
```

另外我们还可以使用分词器，如 OpenNLP 的 WhitespaceTokenizer 类，如下所示：

```
String words[] =
    WhitespaceTokenizer.INSTANCE.tokenize(sentence);
```

5.2.1　使用 OpenNLP 词性标注器

OpenNLP 提供了多个支持词性标注的类。我们将演示如何使用 POSTaggerME 类来执行基本的标注，以及使用 ChunkerME 类来执行分块。分块涉及根据单词的类型进行相关的单词分组。这可以更深入地了解句子的结构。我们还将研究 POSDictionary 实例的创建和使用。

5.2.1.1　使用 OpenNLP 的 POSTaggerME 类词性标注器

OpenNLP 的 POSTaggerME 类使用最大熵原理来进行词性标注。根据这个词本身和这个词的上下文来确定标注类型。任何单词都可能有多个相关的标注。标注器使用概率模型来确定要分配的标注。

从文件加载词性标注模型。en-pos-maxent.bin 模型经常使用，其是基于 Penn TreeBank 标注库。可以在 http://opennlp.sourceforge.net/models-1.5/ 找到 OpenNLP 各种预先训练好的词性标注模型。

加载模块时可能会产生 IOException 异常，所以使用 try-catch 块来加载模型，如下所示，我们使用 en-pos-maxent.bin 来加载模型：

```
try (InputStream modelIn = new FileInputStream(
    new File(getModelDir(), "en-pos-maxent.bin"));) {
    …
}
catch (IOException e) {
    // Handle exceptions
}
```

接下来，创建 POSModel 和 POSTaggerME 实例：

```
POSModel model = new POSModel(modelIn);
POSTaggerME tagger = new POSTaggerME(model);
```

tag 方法使用要处理的文本作为它的参数：

```
String tags[] = tagger.tag(sentence);
```

显示每一个单词和它们的标注：

```
for (int i = 0; i<sentence.length; i++) {
    System.out.print(sentence[i] + "/" + tags[i] + " ");
}
```

结果如下。每个单词后面跟其类型:

The/DT voyage/NN of/IN the/DT Abraham/NNP Lincoln/NNP was/VBD for/IN a/DT long/JJ time/NN marked/VBN by/IN no/DT special/JJ incident./NN

对于任何一句话,每一个单词都可能被分配多个标注。topKSequences 方法将返回一组基于其正确概率的集合。在接下来的代码中,使用 topKSequences 方法处理 sentence 变量的内容并显示结果:

```
Sequence topSequences[] = tagger.topKSequences(sentence);
for (inti = 0; i<topSequences.length; i++) {
    System.out.println(topSequences[i]);
}
```

结果如下,其中的第一个数字表示加权分数,括号内的标注根据分数排列:

-0.5563571615737618 [DT, NN, IN, DT, NNP, NNP, VBD, IN, DT, JJ, NN, VBN, IN, DT, JJ, NN]
-2.9886144610050907 [DT, NN, IN, DT, NNP, NNP, VBD, IN, DT, JJ, NN, VBN, IN, DT, JJ, .]
-3.771930515521527 [DT, NN, IN, DT, NNP, NNP, VBD, IN, DT, JJ, NN, VBN, IN, DT, NN, NN]

 请确保你引用了正确的 Sequence 类。在这个例子中,使用 import opennlp.tools.util.Sequence;。

Sequence 类有多种方法,详见下表:

方 法	含 义
getOutcomes	返回字符串列表其表示句子中每个单词的标注
getProbs	返回 double 类型数组变量其表示序列中为每个标注的概率
getScore	返回标注序列的加权值

在下面的代码中,我们使用这几种方法来说明它们是如何使用的。对于每个标注序列,用斜杠分隔标注及其概率:

```
for (int i = 0; i<topSequences.length; i++) {
    List<String> outcomes = topSequences[i].getOutcomes();
    double probabilities[] = topSequences[i].getProbs();
    for (int j = 0; j <outcomes.size(); j++) {
```

```
            System.out.printf("%s/%5.3f ",outcomes.get(j),
                probabilities[j]);
        }
        System.out.println();
    }
    System.out.println();
```

结果如下。相邻的几行代表一个标注序列:

```
DT/0.992 NN/0.990 IN/0.989 DT/0.990 NNP/0.996 NNP/0.991 VBD/0.994
IN/0.996 DT/0.996 JJ/0.991 NN/0.994 VBN/0.860 IN/0.985 DT/0.960 JJ/0.919
NN/0.832
DT/0.992 NN/0.990 IN/0.989 DT/0.990 NNP/0.996 NNP/0.991 VBD/0.994
IN/0.996 DT/0.996 JJ/0.991 NN/0.994 VBN/0.860 IN/0.985 DT/0.960 JJ/0.919
./0.073
DT/0.992 NN/0.990 IN/0.989 DT/0.990 NNP/0.996 NNP/0.991 VBD/0.994
IN/0.996 DT/0.996 JJ/0.991 NN/0.994 VBN/0.860 IN/0.985 DT/0.960 NN/0.073
NN/0.419
```

5.2.1.2 使用 OpenNLP 分块

分块的过程包括把一个句子分成几个部分或数据块。然后可以用标注注释这些数据块。我们将使用 ChunkerME 类来说明这是如何实现的。这个类会加载一个模型到一个 ChunkerModel 实例。ChunkerME 类的 chunk 方法执行实际分块处理。我们也将研究使用 chunkAsSpans 方法来返回这些分块的信息。这能让我们了解一个分块有多长以及分块是由什么元素组成的。

我们将使用 en-pos-maxent.bin 文件创建一个 POSTaggerME 类实例模型。我们需要使用这个实例来为文本标注,就像上一节一样。我们还将使用 en-chunker.bin 文件创建一个 ChunkerModel 的实例与 ChunkerME 的实例一起使用。

这些模型是通过输入流创建的,如下面的例子所示。

我们使用 try-with-resources 的方式打开和关闭文件,并处理可能抛出的异常:

```
try (
        InputStream posModelStream = new FileInputStream(
            getModelDir() + "\\en-pos-maxent.bin");
        InputStream chunkerStream = new FileInputStream(
            getModelDir() + "\\en-chunker.bin");) {
    …
} catch (IOException ex) {
    // Handle exceptions
}
```

下面的代码创建并使用标注器来找出句子中每个单词的词性。然后显示这句话的单词及其标注：

```
POSModel model = new POSModel(posModelStream);
POSTaggerME tagger = new POSTaggerME(model);

String tags[] = tagger.tag(sentence);
for(int i=0; i<tags.length; i++) {
    System.out.print(sentence[i] + "/" + tags[i] + " ");
}
System.out.println();
```

结果如下。结果显示出来之后也就清楚了解分块器是如何工作的：

```
The/DT voyage/NN of/IN the/DT Abraham/NNP Lincoln/NNP was/VBD for/IN a/DT
long/JJ time/NN marked/VBN by/IN no/DT special/JJ incident./NN
```

使用输入流创建一个 ChunkerModel 实例。然后创建 ChunkerME 实例，紧接着使用其 chunk 方法，如下所示。chunk 方法将使用这句话的词项及其标注来创建一个字符串数组。每个字符串内都包括词项和分块的信息：

```
ChunkerModel chunkerModel = new
    ChunkerModel(chunkerStream);
ChunkerME chunkerME = new ChunkerME(chunkerModel);
String result[] = chunkerME.chunk(sentence, tags);
```

显示结果数组里的每个词项及其分块标注：

```
for (int i = 0; i < result.length; i++) {
    System.out.println("[" + sentence[i] + "] " + result[i]);
}
```

结果如下。方括号中显示句子的词项，其后面跟着其分块标注。下表中介绍了这些标注：

	第一部分
B	标注的开始
I	标注的中间
E	标注的结束（如果分块只包括一个单词，不出现这个结束标注）
	第二部分
NP	名词块
VB	动词块

多个词组成的词组被分在一起，如"The voyage"和"the Abraham Lincoln"。

```
[The] B-NP
[voyage] I-NP
```

```
[of] B-PP
[the] B-NP
[Abraham] I-NP
[Lincoln] I-NP
[was] B-VP
[for] B-PP
[a] B-NP
[long] I-NP
[time] I-NP
[marked] B-VP
[by] B-PP
[no] B-NP
[special] I-NP
[incident.] I-NP
```

如果我们有兴趣获取分块的更多详细信息,我们可以使用 ChunkerME 类的 chunkAsSpans 方法。这个方法返回 Span 对象的数组。每个对象表示文本中的一个 span。

ChunkerME 类还有其他几种方法。在这里,我们将说明如何使用 getType、getStart 和 getEnd 方法。getType 方法返回分块标注的第二部分,getStart 和 getEnd 方法则返回词项在原始句子中开始和结束的索引。length 方法返回词项数量的 span 的长度。

在下面的例子中,使用 sentence 和 tag 数组作为参数执行 chunkAsSpans 方法。然后显示 span 数组。外部循环一次循环处理一个 Span 对象,并显示其基本信息。内部的循环把 span 的文本信息放在括号内显示:

```
Span[] spans = chunkerME.chunkAsSpans(sentence, tags);
for (Span span : spans) {
    System.out.print("Type: " + span.getType() + " - "
        + " Begin: " + span.getStart()
        + " End:" + span.getEnd()
        + " Length: " + span.length() + "  [");
    for (int j = span.getStart(); j < span.getEnd(); j++) {
        System.out.print(sentence[j] + " ");
    }
    System.out.println("]");
}
```

下面的结果清楚地显示 span 风格、在 sentence 数组的位置、长度,以及实际的文本:

```
Type: NP -  Begin: 0 End:2 Length: 2   [The voyage ]
Type: PP -  Begin: 2 End:3 Length: 1   [of ]
```

```
Type: NP  -  Begin: 3  End:6  Length: 3   [the Abraham Lincoln ]
Type: VP  -  Begin: 6  End:7  Length: 1   [was ]
Type: PP  -  Begin: 7  End:8  Length: 1   [for ]
Type: NP  -  Begin: 8  End:11 Length: 3   [a long time ]
Type: VP  -  Begin: 11 End:12 Length: 1   [marked ]
Type: PP  -  Begin: 12 End:13 Length: 1   [by ]
Type: NP  -  Begin: 13 End:16 Length: 3   [no special incident. ]
```

5.2.1.3 使用 POSDictionary 类

标注字典包含每个单词的有效标注。这样可以防止标注被不适当地应用于某个单词。另外，一些搜索算法执行速度更快的原因是因为它们不需要考虑那些出现概率比较少的标注。

在本节中，我们将演示如何：

- 获得标注器的标注词典
- 确定一个词有什么标注
- 更改一个单词的标注
- 添加一个新标注字典到一个新的标注器工厂

与前面的示例一样，我们将使用一个 try-with-resources 块来打开词性标注模型的输入流，然后创建模型和标注器工厂，如下所示：

```
try (InputStream modelIn = new FileInputStream(
        new File(getModelDir(), "en-pos-maxent.bin"));) {
    POSModel model = new POSModel(modelIn);
    POSTaggerFactory posTaggerFactory = model.getFactory();
    …
} catch (IOException e) {
    //Handle exceptions
}
```

获取用于标注的标注词典

我们使用 POSModel 类的 getFactory 方法来得到一个 POSTaggerFactory 实例。然后使用其 getTagDictionary 方法来获取其 TagDictionary 实例。如下所示：

```
MutableTagDictionary tagDictionary =
  (MutableTagDictionary)posTaggerFactory.getTagDictionary();
```

MutableTagDictionary 接口继承自 TagDictionary 接口。TagDictionary 接口有一个 getTags

方法，MutableTagDictionary 接口多了个 put 方法，它允许添加标注到字典中去。这些接口是由 POSDictionary 类实现的。

确定一个单词的标注

若要获取给定单词的标注，请使用 getTags 方法。它用字符串数组的形式返回标注数组。这些标注如下所示：

```
String tags[] = tagDictionary.getTags("force");
for (String tag : tags) {
    System.out.print("/" + tag);
}
System.out.println();
```

结果如下：

/NN/VBP/VB

这意味着能以三种不同的方式解释"force"一词。

更改一个单词的标注

MutableTagDictionary 接口的 put 方法允许我们为一个单词添加一个新标注。该方法具有两个参数：这个单词及其新的标注。然后方法返回一个包含旧标注数组。

在以下示例中，我们使用新标注来替换旧标注。然后显示旧的标注。

```
String oldTags[] = tagDictionary.put("force", "newTag");
for (String tag : oldTags) {
    System.out.print("/" + tag);
}
System.out.println();
```

下面的结果列出这个单词的旧标注：

/NN/VBP/VB

这些标注已被替换为新的标注，如下所示显示当前标注：

```
tags = tagDictionary.getTags("force");
for (String tag : tags) {
    System.out.print("/" + tag);
}
System.out.println();
```

我们得到以下结果：

/newTag

要保留旧的标注，我们需要创建一个字符串数组来保存旧的和新的标注，然后使用该数组作为 put 方法第二个参数，如下所示：

```
String newTags[] = new String[tags.length+1];
for (int i=0; i<tags.length; i++) {
    newTags[i] = tags[i];
}
newTags[tags.length] = "newTag";
oldTags = tagDictionary.put("force", newTags);
```

如果我们要重新显示单词当前标注，可以看到保留了旧标注并添加了新标注，如下所示：

/NN/VBP/VB/newTag

当添加标注时，小心并适当分配标注的顺序，因为顺序会影响分配标注。

添加新的标注字典

新的标注字典可以被添加到一个 POSTaggerFactory 实例。我们将通过创建新的 POSTaggerFactory，然后添加我们之前开发的 tagDictionary 说明这一过程。如下所示，首先使用默认构造器创建一个新的工厂类，然后调用新工厂类的 setTagDictionary 方法。

```
POSTaggerFactory newFactory = new POSTaggerFactory();
newFactory.setTagDictionary(tagDictionary);
```

为了确认已添加的标注字典，我们显示"force"一词的标注如下所示：

```
tags = newFactory.getTagDictionary().getTags("force");
for (String tag : tags) {
    System.out.print("/" + tag);
}
System.out.println();
```

标注是相同的，如下所示：

/NN/VBP/VB/newTag

从文件创建一个字典

如果我们要创建一个新字典，一个方法是创建一个包含单词和它们标注的 XML 文件，然后使用该文件创建字典。OpenNLP 的 POSDictionary 类的 create 方法支持这种字典创建方法。

XML 文件由词典 root 元素后跟一系列 entry 元素构成。entry 元素使用 tags 属性指定单词的标注。单词被包含在 entry 元素作为 token 元素。以下是一个简单的例子，文件 dictionary.txt 为包含两个单词的标注字典：

```xml
<dictionary case_sensitive="false">
    <entry tags="JJ VB">
        <token>strong</token>
    </entry>
    <entry tags="NN VBP VB">
        <token>force</token>
    </entry>
</dictionary>
```

我们基于输入流用 create 方法创建字典，如下所示：

```java
try (InputStream dictionaryIn =
    new FileInputStream(new File("dictionary.txt"));) {
    POSDictionary dictionary =
    POSDictionary.create(dictionaryIn);
    …
} catch (IOException e) {
    // Handle exceptions
}
```

POSDictionary 类有一个 iterator 方法返回一个字符类型的 iterator 对象。其 next 方法返回字典中的每个单词。我们可以使用这些方法来显示字典中的所有内容，如下所示：

```java
Iterator<String> iterator = dictionary.iterator();
while (iterator.hasNext()) {
    String entry = iterator.next();
    String tags[] = dictionary.getTags(entry);
    System.out.print(entry + " ");
    for (String tag : tags) {
        System.out.print("/" + tag);
    }
    System.out.println();
}
```

下面显示了我们预期的结果：

strong /JJ/VB
force /NN/VBP/VB

5.2.2 使用 Stanford 词性标注器

在本节中，我们将研究 Stanford API 支持的两种不同的标注方法。第一种方法使用 MaxentTagger 类。正如其名，它使用最大熵原理来进行词性标注。我们也将使用这个类来设计一个模型处理"短信文"类型的文本。第二种方法将使用流水线和注释器。英文标注器使用 Penn Treebank 英语词类标注库。

5.2.2.1 使用 Stanford MaxentTagger 类

MaxentTagger 类使用模型来执行标注任务。大量的模型及其捆绑的 API 都被包括在 extension.tagger 文件里。它们包括英语、汉语、阿拉伯语、法语和德语的模型。这里列出了英语的模型。模型名称里有 wsj 前缀的，是指基于《华尔街日报》产生的模型。其他词汇指用于训练模型的方法。这些概念在这里不讨论：

- wsj-0-18-bidirectional-distsim.tagger
- wsj-0-18-bidirectional-nodistsim.tagger
- wsj-0-18-caseless-left3words-distsim.tagger
- wsj-0-18-left3words-distsim.tagger
- wsj-0-18-left3words-nodistsim.tagger
- english-bidirectional-distsim.tagger
- english-caseless-left3words-distsim.tagger
- english-left3words-distsim.tagger

本例将从一个文件中读取一系列句子。然后对每个句子进行处理，说明各种访问，并显示单词及其标注的方法。

我们首先使用 try-with-resources 块来处理 IO 异常，如下所示。使用 wsj-0-18-bidirectional-distsim.tagger 文件创建 MaxentTagger 类的一个实例。

使用 MaxentTagger 类的 tokenizeText 方法创建一个接着一个 HasWord 对象的 List 实例。这些句子都从文件 sentences.txt 读入。HasWord 接口代表单词和内容两种方法：setWord 和 word 方法。后一种方法以字符串形式返回一个单词。每个句子由 HasWord 对象的 List 实例来表示：

```
try {
    MaxentTagger tagger = new MaxentTagger(getModelDir() +
        "//wsj-0-18-bidirectional-distsim.tagger");
    List<List<HasWord>> sentences = MaxentTagger.tokenizeText(
        new BufferedReader(new FileReader("sentences.txt")));
```

```
    …
} catch (FileNotFoundException ex) {
    // Handle exceptions
}
```

sentences.txt 文件内容为《海底两万里》这本书第 5 章中的前四句话：

```
The voyage of the Abraham Lincoln was for a long time marked by no
special incident.
But one circumstance happened which showed the wonderful dexterity of
Ned Land, and proved what confidence we might place in him.
The 30th of June, the frigate spoke some American whalers, from whom
we learned that they knew nothing about the narwhal.
But one of them, the captain of the Monroe, knowing that Ned Land had
shipped on board the Abraham Lincoln, begged for his help in chasing a
whale they had in sight.
```

添加一个循环来处理 sentences 列表里的每个句子。tagSentence 方法返回一个 TaggedWord 对象的 List 实例，如下所示。TaggedWord 类实现自 HasWord 接口，并添加一个 tag 方法返回与这个单词相关的标注。如下所示，toString 方法被用来显示每个句子：

```
List<TaggedWord> taggedSentence =
    tagger.tagSentence(sentence);
for (List<HasWord> sentence : sentences) {
    List<TaggedWord> taggedSentence=
        tagger.tagSentence(sentence);
    System.out.println(taggedSentence);
}
```

结果如下：

```
[The/DT, voyage/NN, of/IN, the/DT, Abraham/NNP, Lincoln/NNP, was/VBD,
for/IN, a/DT, long/JJ, --- time/NN, marked/VBN, by/IN, no/DT, special/JJ,
incident/NN, ./.]
 [But/CC, one/CD, circumstance/NN, happened/VBD, which/WDT, showed/VBD,
the/DT, wonderful/JJ, dexterity/NN, of/IN, Ned/NNP, Land/NNP, ,/,, and/
CC, proved/VBD, what/WP, confidence/NN, we/PRP, might/MD, place/VB, in/
IN, him/PRP, ./.]
[The/DT, 30th/JJ, of/IN, June/NNP, ,/,, the/DT, frigate/NN, spoke/
VBD, some/DT, American/JJ, whalers/NNS, ,/,, from/IN, whom/WP, we/PRP,
learned/VBD, that/IN, they/PRP, knew/VBD, nothing/NN, about/IN, the/DT,
narwhal/NN, ./.]
[But/CC, one/CD, of/IN, them/PRP, ,/,, the/DT, captain/NN, of/IN, the/
DT, Monroe/NNP, ,/,, knowing/VBG, that/IN, Ned/NNP, Land/NNP, had/VBD,
shipped/VBN, on/IN, board/NN, the/DT, Abraham/NNP, Lincoln/NNP, ,/,,
begged/VBN, for/IN, his/PRP$, help/NN, in/IN, chasing/VBG, a/DT, whale/
NN, they/PRP, had/VBD, in/IN, sight/NN, ./.]
```

或者我们使用 Sentence 类的 listToString 方法，将要标注的句子转换为一个简单的字符串对象。当 listToString 方法的第二个参数为 false 时，就是使用 HasWord 的 toString 方法来创建生成的字符串，如下所示：

```
List<TaggedWord> taggedSentence =
    tagger.tagSentence(sentence);
for (List<HasWord> sentence : sentences) {
    List<TaggedWord> taggedSentence=
        tagger.tagSentence(sentence);
    System.out.println(Sentence.listToString(taggedSentence, false));
}
```

这就产生了更加美观的结果：

```
The/DT voyage/NN of/IN the/DT Abraham/NNP Lincoln/NNP was/VBD for/IN a/DT
long/JJ time/NN marked/VBN by/IN no/DT special/JJ incident/NN ./.
But/CC one/CD circumstance/NN happened/VBD which/WDT showed/VBD the/DT
wonderful/JJ dexterity/NN of/IN Ned/NNP Land/NNP ,/, and/CC proved/VBD
what/WP confidence/NN we/PRP might/MD place/VB in/IN him/PRP ./.
The/DT 30th/JJ of/IN June/NNP ,/, the/DT frigate/NN spoke/VBD some/DT
American/JJ whalers/NNS ,/, from/IN whom/WP we/PRP learned/VBD that/IN
they/PRP knew/VBD nothing/NN about/IN the/DT narwhal/NN ./.
But/CC one/CD of/IN them/PRP ,/, the/DT captain/NN of/IN the/DT Monroe/
NNP ,/, knowing/VBG that/IN Ned/NNP Land/NNP had/VBD shipped/VBN on/IN
board/NN the/DT Abraham/NNP Lincoln/NNP ,/, begged/VBN for/IN his/PRP$
help/NN in/IN chasing/VBG a/DT whale/NN they/PRP had/VBD in/IN sight/NN
./.
```

我们可以使用以下代码得到相同的结果。word 和 tag 方法用来获取每个单词及其标注：

```
List<TaggedWord> taggedSentence =
    tagger.tagSentence(sentence);
for (TaggedWord taggedWord : taggedSentence) {
    System.out.print(taggedWord.word() + "/" +
        taggedWord.tag() + " ");
}
System.out.println();
```

如果我们只对获取某一个给定标注感兴趣，可以使用如下方法，下面将只列出单数名词标注（NN）：

```
List<TaggedWord> taggedSentence =
    tagger.tagSentence(sentence);
for (TaggedWord taggedWord : taggedSentence) {
    if (taggedWord.tag().startsWith("NN")) {
        System.out.print(taggedWord.word() + " ");
    }
}
System.out.println();
```

下面代码显示每个句子中有单数名字标注的单词：

```
NN Tagged: voyage Abraham Lincoln time incident
NN Tagged: circumstance dexterity Ned Land confidence
NN Tagged: June frigate whalers nothing narwhal
NN Tagged: captain Monroe Ned Land board Abraham Lincoln help whale sight
```

5.2.2.2 使用 MaxentTagger 类来标注短信文

我们可以使用一个不同的模型来处理可能包括短信文的推特文本。GATE（https://gate.ac.uk/wiki/twitter-postagger.html）是一个已经训练好的用于处理推特文本的模型。该模型在这里用于处理短信文：

```
MaxentTagger tagger = new MaxentTagger(getModelDir()
    + "//gate-EN-twitter.model");
```

在这里，我们使用 MaxentTagger 类的 tagString 方法来处理来自 5.1.2 节中的短信文：

```
System.out.println(tagger.tagString("AFAIK she H8 cth!"));
System.out.println(tagger.tagString(
    "BTW had a GR8 tym at the party BBIAM."));
```

结果如下：

```
AFAIK_NNP she_PRP H8_VBP cth!_NN
BTW_UH had_VBD a_DT GR8_NNP tym_NNP at_IN the_DT party_NN BBIAM._NNP
```

5.2.2.3 使用 Stanford 流水线进行标注

我们已经在前面若干例子中使用了 Stanford 流水线。在这个例子中，我们将使用 Stanford 流水线获取词性标注。正如我们之前使用 Stanford API 的例子中，我们创建一个基于一组注释的流水线：tokenize、ssplit 和 pos。

这些将对文本进行分词，然后分隔成句子，最后获取词性标注：

```
Properties props = new Properties();
props.put("annotators", "tokenize, ssplit, pos");
StanfordCoreNLP pipeline = new StanfordCoreNLP(props);
```

我们将使用 theSentence 变量作为要处理的文字并放到注释里。调用流水线的 annotate 方法，如下所示：

```
Annotation document = new Annotation(theSentence);
pipeline.annotate(document);
```

流水线可以执行不同类型的处理，CoreMap 对象的列表用来访问单词和标注。使用 Annotation 类的 get 方法返回句子的列表，如下所示：

```
List<CoreMap> sentences =
    document.get(SentencesAnnotation.class);
```

可以使用其 get 方法访问 CoreMap 对象的内容。该方法的参数是所需信息的类。下面的代码示例使用 TextAnnotation 类可访问词项，使用 PartOfSpeechAnnotation 类可查找词性标注，并列出了每个句子中的每个单词及其标注：

```
for (CoreMap sentence : sentences) {
    for (CoreLabel token : sentence.get(TokensAnnotation.class)) {
        String word = token.get(TextAnnotation.class);
        String pos = token.get(PartOfSpeechAnnotation.class);
        System.out.print(word + "/" + pos + " ");
    }
    System.out.println();
}
```

结果如下：

```
The/DT voyage/NN of/IN the/DT Abraham/NNP Lincoln/NNP was/VBD for/IN a/DT
long/JJ time/NN marked/VBN by/IN no/DT special/JJ incident/NN ./.
```

流水线可以使用其他选项来控制标注器如何工作。例如，默认情况下使用 english-left3words-distsim.tagger 标注器模型。我们可以使用 pos.model 属性来指定一个不同的模型，如下所示。此外，还有一个 pos.maxlen 属性来控制句子的最大数量：

```
props.put("pos.model",
    "C:/.../Models/english-caseless-left3words-distsim.tagger");
```

有时使用 XML 格式文件是很有用的。可以使用 StanfordCoreNLP 类的 xmlPrint 方法写出这样的 XML 文件。该方法的第一个参数是要显示的注释。它的第二个参数是要写入的 OutputStream 对象。看下面的代码，标注结果将写到标准输出。这里需要使用 try-catch 块来处理 IO 异常：

```
try {
    pipeline.xmlPrint(document, System.out);
} catch (IOException ex) {
    // Handle exceptions
}
```

部分结果如下所示。只列出了前两个单词和最后一个单词。每个词项标注包含这个单词、其位置及其词性标注：

```
<?xml version="1.0" encoding="UTF-8"?>
<?xml-stylesheet href="CoreNLP-to-HTML.xsl" type="text/xsl"?>
<root>
<document>
```

```xml
<sentences>
<sentence id="1">
<tokens>
<token id="1">
<word>The</word>
<CharacterOffsetBegin>0</CharacterOffsetBegin>
<CharacterOffsetEnd>3</CharacterOffsetEnd>
<POS>DT</POS>
</token>
<token id="2">
<word>voyage</word>
<CharacterOffsetBegin>4</CharacterOffsetBegin>
<CharacterOffsetEnd>10</CharacterOffsetEnd>
<POS>NN</POS>
</token>
...
<token id="17">
<word>.</word>
<CharacterOffsetBegin>83</CharacterOffsetBegin>
<CharacterOffsetEnd>84</CharacterOffsetEnd>
<POS>.</POS>
</token>
</tokens>
</sentence>
</sentences>
</document>
</root>
```

prettyPrint 方法与 xmlPrint 方法类似：

```
pipeline.prettyPrint(document, System.out);
```

但是输出不是很好看，如下所示。先显示原始句子，其后跟着显示每个单词及其位置和标注。输出已被格式化，使其更具有可读性：

```
The voyage of the Abraham Lincoln was for a long time marked by no
special incident.
[Text=The CharacterOffsetBegin=0 CharacterOffsetEnd=3 PartOfSpeech=DT]
[Text=voyage CharacterOffsetBegin=4 CharacterOffsetEnd=10
PartOfSpeech=NN]
[Text=of CharacterOffsetBegin=11 CharacterOffsetEnd=13 PartOfSpeech=IN]
[Text=the CharacterOffsetBegin=14 CharacterOffsetEnd=17 PartOfSpeech=DT]
```

```
[Text=Abraham CharacterOffsetBegin=18 CharacterOffsetEnd=25
PartOfSpeech=NNP]
 [Text=Lincoln CharacterOffsetBegin=26 CharacterOffsetEnd=33
PartOfSpeech=NNP]
 [Text=was CharacterOffsetBegin=34 CharacterOffsetEnd=37
PartOfSpeech=VBD]
 [Text=for CharacterOffsetBegin=38 CharacterOffsetEnd=41 PartOfSpeech=IN]
 [Text=a CharacterOffsetBegin=42 CharacterOffsetEnd=43 PartOfSpeech=DT]
 [Text=long CharacterOffsetBegin=44 CharacterOffsetEnd=48
PartOfSpeech=JJ]
 [Text=time CharacterOffsetBegin=49 CharacterOffsetEnd=53
PartOfSpeech=NN]
 [Text=marked CharacterOffsetBegin=54 CharacterOffsetEnd=60
PartOfSpeech=VBN]
 [Text=by CharacterOffsetBegin=61 CharacterOffsetEnd=63 PartOfSpeech=IN]
 [Text=no CharacterOffsetBegin=64 CharacterOffsetEnd=66 PartOfSpeech=DT]
 [Text=special CharacterOffsetBegin=67 CharacterOffsetEnd=74
PartOfSpeech=JJ]
 [Text=incident CharacterOffsetBegin=75 CharacterOffsetEnd=83
PartOfSpeech=NN]
 [Text=. CharacterOffsetBegin=83 CharacterOffsetEnd=84 PartOfSpeech=.]
```

5.2.3 使用 LingPipe 词性标注器

LingPipe 使用 Tagger 接口来支持词性标注。这个接口只有一个方法：tag。它返回 Tagging 对象的 List 实例。这些对象包含单词及其标注。接口由 ChainCrf 和 HmmDecoder 类实现。

ChainCrf 类使用马尔可夫链的条件随机场解码和估计来确定标注。HmmDecoder 类使用隐马尔可夫模型来进行标注。接下来我们将说明这个类。

HmmDecoder 类使用 tag 方法来判断最有可能的（最优的）标注。它还有一个 tagNBest 方法给这些可能的标注打分，然后返回这些被打分标注的迭代器。有三种依赖于 LingPipe 的词性标注模型，可以从 http://alias-i.com/lingpipe/web/models.html 下载。这些在下表中列出。对于我们的示例，将使用 Brown 语料库模式：

模　型	文　件
英语常用词：Brown 语料库	pos-en-general-brown. HiddenMarkovModel
英语生物医学词汇：MedPost 语料库	pos-en-bio-medpost. HiddenMarkovModel
英语生物医学词汇：GENIA 语料库	pos-en-bio-genia. HiddenMarkovModel

5.2.3.1 使用 HmmDecoder 类的 Best_First 标注

我们使用 try-with-resources 块来处理异常和创建 HmmDecoder 实例代码,如下所示。模型信息是从文件中读取,然后作为 HmmDecoder 构造器的参数:

```
try (
        FileInputStream inputStream =
            new FileInputStream(getModelDir()
            + "//pos-en-general-brown.HiddenMarkovModel");
        ObjectInputStream objectStream =
            new ObjectInputStream(inputStream);) {
    HiddenMarkovModel hmm = (HiddenMarkovModel)
        objectStream.readObject();
    HmmDecoder decoder = new HmmDecoder(hmm);
    …
} catch (IOException ex) {
 // Handle exceptions
} catch (ClassNotFoundException ex) {
 // Handle exceptions
};
```

我们将对 theSentence 变量进行标注。首先,使用印欧语系分词器进行分词,如下所示。tokenizer 方法需要把文本字符串转换为字符数组,然后 tokenize 方法返回词项字符串的数组:

```
TokenizerFactory TOKENIZER_FACTORY =
    IndoEuropeanTokenizerFactory.INSTANCE;
char[] charArray = theSentence.toCharArray();
Tokenizer tokenizer =
    TOKENIZER_FACTORY.tokenizer(
        charArray, 0, charArray.length);
String[] tokens = tokenizer.tokenize();
```

实际的标注工作是由 HmmDecoder 类的 tag 方法进行。不过,这个方法需要一个 String 词项的 List 实例。这个列表可以使用 Arrays 类的 asList 方法来创建。Tagging 类包含词项和标注的序列:

```
List<String> tokenList = Arrays.asList(tokens);
Tagging<String> tagString = decoder.tag(tokenList);
```

现在我们已经准备好列出词项及其标注。下面的循环使用 token 和 tag 方法来访问词项和标注,并分别添加在 Tagging 对象中。代码如下:

```
for (int i = 0; i < tagString.size(); ++i) {
    System.out.print(tagString.token(i) + "/"
    + tagString.tag(i) + " ");
}
```

结果如下：

```
The/at voyage/nn of/in the/at Abraham/np Lincoln/np was/bedz for/in a/at
long/jj time/nn marked/vbn by/in no/at special/jj incident/nn ./.
```

5.2.3.2　使用 HmmDecoder 类的 NBest 标注

标注过程需要考虑到标注的多种组合。HmmDecoder 类的 tagNBest 方法返回 ScoredTagging 对象的迭代器，其反映了不同标注置信度的顺序。此方法需要一个词项列表和所需结果的最大数量。

上面那个例句没有歧义，不能说明标注的多种组合。我们将使用下面这句话作为例子：

```
String[] sentence = {"Bill", "used", "the", "force",
    "to", "force", "the", "manager", "to",
    "tear", "the", "bill","in", "to."};
List<String> tokenList = Arrays.asList(sentence);
```

使用此方法时，需要在开始时声明一个变量表示结果数量：

```
int maxResults = 5;
```

我们使用上一节中创建的 decoder 对象，对其使用 tagNBest 方法，如下所示：

```
Iterator<ScoredTagging<String>> iterator =
    decoder.tagNBest(tokenList, maxResults);
```

迭代器允许获取每 5 个不同的分数。ScoredTagging 类有个 score 方法，其返回值可以反映其标注的好坏。在下面的代码中，使用 printf 语句显示这个分数。后面跟着一个循环列出词项及其标注。显示的结果是一个标注的得分及带有标注的单词序列：

```
while (iterator.hasNext()) {
    ScoredTagging<String> scoredTagging = iterator.next();
    System.out.printf("Score: %7.3f   Sequence: ",
        scoredTagging.score());
    for (int i = 0; i < tokenList.size(); ++i) {
        System.out.print(scoredTagging.token(i) + "/"
            + scoredTagging.tag(i) + " ");
    }
    System.out.println();
}
```

结果如下。请注意，单词"force"可以有 nn、jj 或 vb 三种标注：

```
Score: -148.796   Sequence: Bill/np used/vbd the/at force/nn to/to force/
vb the/at manager/nn to/to tear/vb the/at bill/nn in/in two./nn

Score: -154.434   Sequence: Bill/np used/vbn the/at force/nn to/to force/
vb the/at manager/nn to/to tear/vb the/at bill/nn in/in two./nn
```

```
Score: -154.781    Sequence: Bill/np used/vbd the/at force/nn to/in force/
nn the/at manager/nn to/to tear/vb the/at bill/nn in/in two./nn
Score: -157.126    Sequence: Bill/np used/vbd the/at force/nn to/to force/
vb the/at manager/jj to/to tear/vb the/at bill/nn in/in two./nn
Score: -157.340    Sequence: Bill/np used/vbd the/at force/jj to/to force/
vb the/at manager/nn to/to tear/vb the/at bill/nn in/in two./nn
```

5.2.3.3 使用 HmmDecoder 类判断标注的置信度

可利用格子结构进行备选单词顺序的统计分析。这种结构代表其向前/向后的分数。HmmDecoder 类的 tagMarginal 方法返回一个 TagLattice 类的实例,它代表一个格子。

我们可以使用 ConditionalClassification 类的实例检查每个格子的词项。在以下示例中,tagMarginal 方法返回一个 TagLattice 实例。循环用于获取每个格子的词项的 ConditionalClassification 实例。

我们使用在上一节中开发的同一个 tokenList 实例:

```
TagLattice<String> lattice = decoder.tagMarginal(tokenList);
for (int index = 0; index < tokenList.size(); index++) {
    ConditionalClassification classification =
        lattice.tokenClassification(index);
    …
}
```

ConditionalClassification 类有 score 方法和 category 方法。score 方法返回给定类别的相对分值。category 方法返回这个标注的类别。词项的得分和类别如下所示:

```
System.out.printf("%-8s",tokenList.get(index));
for (int i = 0; i < 4; ++i) {
    double score = classification.score(i);
    String tag = classification.category(i);
    System.out.printf("%7.3f/%-3s ",score,tag);
}
System.out.println();
```

输出结果如下所示:

```
Bill     0.974/np    0.018/nn    0.006/rb    0.001/nps
used     0.935/vbd   0.065/vbn   0.000/jj    0.000/rb
the      1.000/at    0.000/jj    0.000/pps   0.000/pp$$
force    0.977/nn    0.016/jj    0.006/vb    0.001/rb
to       0.944/to    0.055/in    0.000/rb    0.000/nn
force    0.945/vb    0.053/nn    0.002/rb    0.001/jj
the      1.000/at    0.000/jj    0.000/vb    0.000/nn
```

manager	0.982/nn	0.018/jj	0.000/nn$	0.000/vb
to	0.988/to	0.012/in	0.000/rb	0.000/nn
tear	0.991/vb	0.007/nn	0.001/rb	0.001/jj
the	1.000/at	0.000/jj	0.000/vb	0.000/nn
bill	0.994/nn	0.003/jj	0.002/rb	0.001/nns
in	0.990/in	0.004/rp	0.002/nn	0.001/jj
two.	0.960/nn	0.013/np	0.011/nns	0.008/rb

5.2.4　训练 OpenNLP 词性标注模型

训练一个 OpenNLP POSModel 的过程与前面训练示例类似。这需要一个训练文件且要足够大，并且是一个高质量的样本集。在训练文件中的每一个语句都必须在一行里，每行由一个词项、字符下划线及其标注组成。

下面的训练数据是小说《海底两万里》第 5 章中的前五句话。虽然这一样本集并不大，但很容易创建并且足以满足我们的目的。

它被保存在名为 sample.train 的文件中：

```
The_DT voyage_NN of_IN the_DT Abraham_NNP Lincoln_NNP was_VBD for_IN a_DT long_JJ time_NN marked_VBN by_IN no_DT special_JJ incident._NN
But_CC one_CD circumstance_NN happened_VBD which_WDT showed_VBD the_DT wonderful_JJ dexterity_NN of_IN Ned_NNP Land,_NNP and_CC proved_VBD what_WP confidence_NN we_PRP might_MD place_VB in_IN him._PRP$
The_DT 30th_JJ of_IN June,_NNP the_DT frigate_NN spoke_VBD some_DT American_NNP whalers,_, from_IN whom_WP we_PRP learned_VBD that_IN they_PRP knew_VBD nothing_NN about_IN the_DT narwhal._NN
But_CC one_CD of_IN them,_PRP$ the_DT captain_NN of_IN the_DT Monroe,_NNP knowing_VBG that_IN Ned_NNP Land_NNP had_VBD shipped_VBN on_IN board_NN the_DT Abraham_NNP Lincoln,_NNP begged_VBD for_IN his_PRP$ help_NN in_IN chasing_VBG a_DT whale_NN they_PRP had_VBD in_IN sight._NN
```

我们将演示如何使用 POSModel 类的 train 方法创建模型，以及如何将模型保存到一个文件中。我们先从声明 POSModel 类的实例变量开始：

```
POSModel model = null;
```

用 try-with-resources 块打开示例文件：

```
try (InputStream dataIn = new FileInputStream("sample.train");) {
    …
} catch (IOException e) {
    // Handle excpetions
}
```

创建 PlainTextByLineStream 类的实例,并与 WordTagSampleStream 类用于创建 ObjectStream<POSSample> 实例。这是 train 方法所需样本数据的格式:

```
ObjectStream<String> lineStream =
    new PlainTextByLineStream(dataIn, "UTF-8");
ObjectStream<POSSample> sampleStream =
    new WordTagSampleStream(lineStream);
```

train 方法的参数为样本所属语言、样本数据字节流、训练参数,以及所需字典(可以为空),如下所示:

```
model = POSTaggerME.train("en", sampleStream,
    TrainingParameters.defaultParams(), null, null);
```

此过程的输出是冗长的。为了节省空间,下面的输出已有所省略:

```
Indexing events using cutoff of 5

  Computing event counts...  done. 90 events
  Indexing... done.
Sorting and merging events... done. Reduced 90 events to 82.
Done indexing.
Incorporating indexed data for training...
done.
  Number of Event Tokens: 82
      Number of Outcomes: 17
      Number of Predicates: 45
...done.
Computing model parameters ...
Performing 100 iterations.
  1:  ... loglikelihood=-254.98920096505964   0.14444444444444443
  2:  ... loglikelihood=-201.19283975630537   0.6
  3:  ... loglikelihood=-174.8849213436524    0.6111111111111112
  4:  ... loglikelihood=-157.58164262220754   0.6333333333333333
  5:  ... loglikelihood=-144.69272379986646   0.6555555555555556
...
  99: ... loglikelihood=-33.461128002846024   0.9333333333333333
 100: ... loglikelihood=-33.29073273669207    0.9333333333333333
```

若要将模型保存到一个文件,我们使用下面的代码。创建输出流并使用 POSModel 类的 serialize 方法保存模型到 en_pos_verne.bin 文件中:

```
try (OutputStream modelOut = new BufferedOutputStream(
    new FileOutputStream(new File("en_pos_verne.bin")));) {
```

```
        model.serialize(modelOut);
} catch (IOException e) {
    // Handle exceptions
}
```

5.3 本章小结

词性标注是一个强大的技术，用于标注一个句子的语法成分。它为下游任务提供了有用的信息，如问题分析和文本情感分析。在第 7 章中，我们仍将讨论这一话题。

由于大多数语言中都有歧义性，标注不是一个容易的过程，尤其是越来越多地使用缩略语使这个过程更加困难。幸运的是，有模型可以很好地做到识别这种类型的文本。然而，随着新的术语和俚语出现，这些模型必须不断更新。

我们研究了 OpenNLP、Stanford API 和 LingPipe 所支持的词性标注功能。这些库使用了几种不同类型的方法（包括基于规则和基于模型的方法）进行词性标注。我们看到使用字典可以提高标注效果。

我们简要介绍了模型的训练过程。预标注过的示例文本被输入并处理后，再形成模型作为输出。虽然我们没有涉及该模型的验证，但可参照前面的章节完成对其的验证。

各种词性标注的方法可以基于许多因素进行比较，比如其准确性和运行速度。虽然本章没有涉及这些问题，但有众多的网络资源可以使用。可以在 http://mattwilkens.com/2008/11/08/evaluating-pos-taggers-speed/ 找到对它们运行速度的比较研究。

在下一章中，我们将研究基于文档内容的分类方法。

CHAPTER 6

第 6 章

文 本 分 类

本章将介绍如何使用各种 NLP 的应用程序接口（API）来进行文本分类。与文本分类不同，文本聚类进行识别时是不依靠预先定义的类别的。相反，文本分类需要预先定义类别，它使用标注来表明文本的类型，这将是本章讨论的重点。

对于文本分类问题，通常的做法是首先训练一个模型，然后对该模型进行验证，之后使用模型进行文档分类。本章我们把重点放在训练模型和使用模型上。

文档可以根据它的多种属性划分成不同的类别，比如文档主题、文档类型、发布时间、作者、使用的语言以及阅读级别等。某些分类方法需要人工对样本数据进行标注。

情感分析是文本分类的一种，它可以判断作者通过哪段文字向读者表明其持有的是正面态度还是负面态度。本章中，我们也会探讨相关技术来实现此类分析。

6.1 文本分类问题

文本分类可用于多种用途：

- 垃圾邮件检测
- 作者身份识别
- 情感分析
- 年龄、性别识别
- 文档主题识别
- 语言识别

对于多数用户来说，垃圾邮件是令人头痛的问题。如果一封电子邮件被划分为垃圾邮件，它将会被移到垃圾邮件目录中。通过分析一段文本消息并根据某些属性可以判断该邮件是否为垃圾邮件。这些属性可以是错误的拼写、不正确的邮件地址和非标准的 URL。

文本分类还可以用来识别作者的身份。这已经被广泛用于识别历史文档的作者，比如《联邦论》（*The Federalist Papers*）和《原色》（*Primary Colors*）的作者识别。

情感分析是一项可以识别文本中作者态度的技术。情感分析在影评领域已经非常流行，除此之外，它也可用于分析几乎所有的产品评价内容，这可以帮助企业更好地评估它们的产品。通常文本中包含正面或负面的属性。情感分析又称为观点提取、观点挖掘或主观性分析。股市的消费者信心以及表现都可以通过推特信息来源或其他资源进行预测。

文本分类还可以用来推测作者的年龄、性别，并能挖掘关于作者更多的信息。代词、限定词和名词短语的数量经常被用于判断作者的性别。女性倾向于使用更多的代词，而男性倾向于使用更多的限定词。

当我们需要整理大量文档时，判断文本主题是非常有用的。虽然搜索引擎也可以处理类似问题，但它只是简单地使用标注云等方法完成文档的归类。标注云是一组关键词的视觉化描述，用于体现每个关键词出现的相对频率。

下图是一个由 IBM 词云生成器（IBM Word Cloud Generator，http://www.softpedia.com/get/Office-tools/Other-Office-Tools/IBM-Word-Cloud-Generator.shtml）创建的标注云

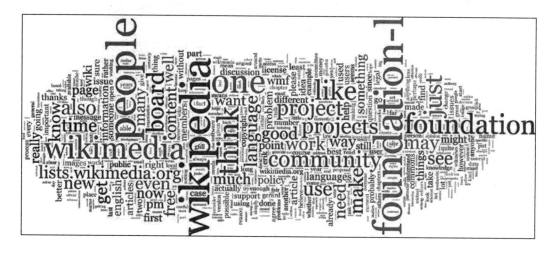

示例。该示例也可通过以下链接获取：https://upload.wikimedia.org/wikipedia/commons/9/9e/Foundation-l_word_cloud_without_headers_and_quotes.png。

文本分类也支持文档语言的识别。许多 NLP 问题是需要应用特定语言模型的，文档语言识别分析对于这类问题是非常有用的。

6.2 情感分析介绍

在情感分析中，我们关注的是某类人群对特定产品或主题持有的态度。通过这项技术，我们可以了解到市民对于当地球队的表现是否满意，他们可能会和球队的管理层持有不同意见。

情感分析可以自动判断消费者对于产品的满意度，然后以更直观的方式展现出来。例如下图所示的是 Kelley Blue Book 网站发布的消费者关于 2014 款凯美瑞（Camry）的评论（http://www.kbb.com/toyota/camry/2014-toyota-camry/?r=471659652516861060）。

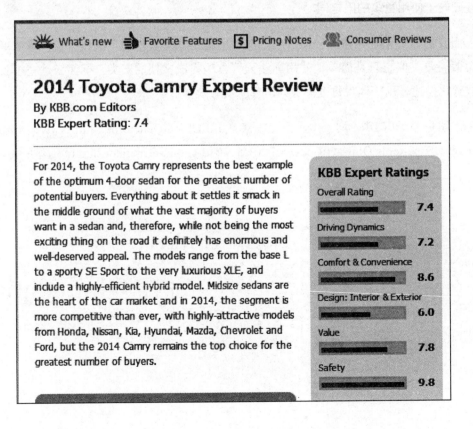

比如像 Overall Rating（综合评分）和 Value（价格）这些属性都是通过柱状图以及数值来展现的。这些数值是可以通过情感分析计算得到的。

情感分析也可用在句子、短语或整篇文档中。它可以通过多种形式呈现：正面、负面，或者 1～10 的数值，甚至可以通过更多复杂的属性类别来呈现。在单句或文档中，更复杂的是对于不同的主题有不同的情感。

如何判断词汇中所包含的情绪？这个问题可以使用情感词典来解决。这里的情感词典是包含不同词语含义的字典。General Inquirer（http://www.wjh.harvard.edu/~inquirer/）就是这样的一个词典，它收录了 1915 个积极情绪词，同时也收录了含有其他属性（如疼痛、愉悦、强烈、动机等）的单词。类似的词汇库还有很多，比如 MPQA Subjectivity Cues Lexicon（http://mpqa.cs.pitt.edu/）。

实践中，我们有时需要创建自己的一个词典，这通常可以通过半监督学习来完成。在此过程中，使用部分标注的样本或者规则来完成词典的创建。这种情形适用于词典和待解决问题不能很好匹配的情况。

我们所关注的不仅是从文本中获取相关情绪，同时也关注相关属性的判断，有时也称为情感对象。让我们来考虑下面这个例子：

"The ride was very rough but the attendants did an excellent job of making us comfortable."

这个句子包含两类情绪：roughness（艰辛）和 comfortable（舒适），前者是负面情绪，后者是正面情绪。正面情绪的对象（或属性）是 job（服务），而负面情绪的对象是 ride（旅程）。

6.3 文本分类技术

文本分类是指给定一个特定文档，判断其是否属于某个类别。文本分类包含两个基本的技术：

- 基于规则的分类
- 监督机器学习

基于规则的分类综合利用词组和其他一些属性，这些属性是根据专家制定的规则组织的，它们对于分类非常有效，但创建过程相当耗时。

监督机器学习（Supervised Machine Learning，SML）使用标注的训练文档创建模型。这个模型通常称为"分类器"。目前有很多不同的机器学习技术，比如朴素贝叶斯（Naïve Bayes）、支持向量机（Support Vector Machine）、k-近邻算法（k-nearest neighbor）等。

本书不会详细介绍这些方法的原理，有兴趣的读者可以参阅相关技术资料。

6.4 使用 API 进行文本分类

本节将通过使用 OpenNLP、Stanford API 和 LingPipe 来演示各种分类方法。我们将着重介绍 LingPipe，因为它提供了多种不同的分类方法。

6.4.1 OpenNLP 的使用

DocumentCategorizer 接口指定了用于分类处理的方法。该接口由 DocumentCategorizerME 类实现，该类使用最大熵模型框架将文本划分到预先定义的类别中。接下来将会介绍：

- 模型的训练
- 模型的使用

6.4.1.1 训练 OpenNLP 分类模型

首先，我们需要训练自己的模型，因为 OpenNLP 不提供预先建立的模型。训练过程首先要创建训练样本文件，然后通过 DocumentCategorizerME 完成实际的训练。由此创建的模型通常被保存到文件中，以便后续使用。

训练样本文件的格式由许多行组成，其中每行代表一个文档。每行的第一个词组表示其所属类别。类别之后由空格隔开一段文本。以下是"狗"（dog）类的一个样本示例：

```
dog The most interesting feature of a dog is its ...
```

为了演示训练过程，我们创建了 en-animals.train 文件，其中包含两个类别："猫"（cat）类和"狗"（dog）类。关于训练文本，我们选取维基百科中的部分内容。对于"狗"（dog）类（https://en.wikipedia.org/wiki/Dog），选取"宠物狗"（As Pets）一节。关于"猫"（cat）

类(http://en.wikipedia.org/wiki/Cats_and_humans),我们选取"宠物"(Pet)一节以及"家养品种"(Domesticated varieties)一节中的第一段描述。我们去掉了文本中的数字型的参考文献。

en-animals.train 文件中每行的部分文字如下:

```
dog The most widespread form of interspecies bonding occurs ...
dog There have been two major trends in the changing status of ...
dog There are a vast range of commodity forms available to ...
dog An Australian Cattle Dog in reindeer antlers sits on Santa's lap
...
dog A pet dog taking part in Christmas traditions ...
dog The majority of contemporary people with dogs describe their ...
dog Another study of dogs' roles in families showed many dogs have
...
dog According to statistics published by the American Pet Products
...
dog The latest study using Magnetic resonance imaging (MRI) ...
cat Cats are common pets in Europe and North America, and their ...
cat Although cat ownership has commonly been associated ...
cat The concept of a cat breed appeared in Britain during ...
cat Cats come in a variety of colors and patterns. These are physical
...
cat A natural behavior in cats is to hook their front claws
periodically ...
cat Although scratching can serve cats to keep their claws from
growing ...
```

在创建训练数据时,使用大量的样本是非常重要的。对于某些分析来说,我们这里使用的数据量是不够的。但是,接下来我们会看到,即使这样,这些数据生成的模型已经能准确地进行分类了。

DoccatModel 类用于文本的归类和分类。根据标注的文本,调用 train 方法训练模型。train 方法有两个参数:第一个参数是 String 类型,用于表示语言,第二个参数是 ObjectStream<DocumentSample> 类型,存放的是训练数据。DocumentSample 实例存放的是标注的文本及其所属类别。

在下面的例子中,使用 en-animal.train 训练模型,它的输入流用来创建一个 PlainTextByLineStream 实例,接着将其转换成 ObjectStream<DocumentSample> 类型,然后调用 train 方法。这段代码需要使用 try-with-resources 来处理异常。另外,我们还创建了输出流用来保存该模型:

```
DoccatModel model = null;
try (InputStream dataIn =
        new FileInputStream("en-animal.train");
     OutputStream dataOut =
        new FileOutputStream("en-animal.model");) {
    ObjectStream<String> lineStream
        = new PlainTextByLineStream(dataIn, "UTF-8");
    ObjectStream<DocumentSample> sampleStream =
        new DocumentSampleStream(lineStream);
    model = DocumentCategorizerME.train("en", sampleStream);
    ...
} catch (IOException e) {
// Handle exceptions
}
```

输出结果如下。为节省空间,这里省略了部分信息:

```
Indexing events using cutoff of 5

  Computing event counts...   done. 12 events
  Indexing...   done.
Sorting and merging events... done. Reduced 12 events to 12.
Done indexing.
Incorporating indexed data for training...
done.
  Number of Event Tokens: 12
      Number of Outcomes: 2
     Number of Predicates: 30
...done.
Computing model parameters ...
Performing 100 iterations.
   1:  ... loglikelihood=-8.317766166719343   0.75
   2:  ... loglikelihood=-7.1439957443937265  0.75
   3:  ... loglikelihood=-6.560690872956419   0.75
   4:  ... loglikelihood=-6.106743124066829   0.75
   5:  ... loglikelihood=-5.721805583104927   0.8333333333333334
   6:  ... loglikelihood=-5.38891508904777785 0.8333333333333334
   7:  ... loglikelihood=-5.098768040466029   0.8333333333333334
...
  98:  ... loglikelihood=-1.4117372921765519  1.0
  99:  ... loglikelihood=-1.4052738190352423  1.0
 100:  ... loglikelihood=-1.398916120150312   1.0
```

这里显示的模型是通过 serialize 方法保存的。其保存为 en-animal.model 文件,可以利用之前提到过的 try-with-resources 块打开：

```
OutputStream modelOut = null;
modelOut = new BufferedOutputStream(dataOut);
model.serialize(modelOut);
```

6.4.1.2　使用 DocumentCategorizerME 进行分类

一旦模型训练完成，我们就可以使用 DocumentCategorizerME 类进行文本分类了。这需要读取模型，创建 DocumentCategorizerME 类实例，然后调用 categorize 方法获取待分类文本所属类别的一组概率值。

由于涉及文件读取，这里需要进行异常处理：

```
try (InputStream modelIn =
        new FileInputStream(new File("en-animal.model"));) {
    ...
} catch (IOException ex) {
    // Handle exceptions
}
```

通过输入流可以创建 DoccatModel 的实例和 DocumentCategorizerME 类，如下所示：

```
DoccatModel model = new DoccatModel(modelIn);
DocumentCategorizerME categorizer =
    new DocumentCategorizerME(model);
```

调用 categorize 方法需要传入一个字符串类型的参数，其返回结果是一组双精度浮点数，每个数值表示待分类文本属于某个类别的似然值。

DocumentCategorizerME 类的 getNumberOfCategories 方法返回模型中类别的个数。DocumentCategorizerME 类的 getCategory 方法根据给定的索引返回所对应的类别。

在下面这段代码中，我们使用了这些方法来显示每个类别及其对应的似然值：

```
double[] outcomes = categorizer.categorize(inputText);
for (int i = 0; i<categorizer.getNumberOfCategories(); i++) {
    String category = categorizer.getCategory(i);
    System.out.println(category + " - " + outcomes[i]);
}
```

在测试阶段，我们从维基百科中选取关于《绿野仙踪》主人公桃乐丝（Dorothy）的狗托托（Toto）的一段文字（https://en.wikipedia.org/wiki/Toto_%28Oz%29）。这里使用"经典书籍"（The classic books）一节中的第一句，声明如下：

```
String toto = "Toto belongs to Dorothy Gale, the heroine of "
     + "the first and many subsequent books. In the first "
     + "book, he never spoke, although other animals, native "
     + "to Oz, did. In subsequent books, other animals "
     + "gained the ability to speak upon reaching Oz or "
     + "similar lands, but Toto remained speechless.";
```

对于"猫"类的测试,我们选取了维基百科文章中"三色猫"(Tortoiseshell and Calico)一节的第一句话(http://en.wikipedia.org/wiki/Cats_and_humans),并声明如下:

```
String calico = "This cat is also known as a calimanco cat or "
     + "clouded tiger cat, and by the abbreviation 'tortie'. "
     + "In the cat fancy, a tortoiseshell cat is patched "
     + "over with red (or its dilute form, cream) and black "
     + "(or its dilute blue) mottled throughout the coat.";
```

通过测试关于 toto 的文本,得到如下输出结果,表明这段文本应该被归类为"狗"类:

```
dog - 0.5870711529777994
cat - 0.41292884702220056
```

而使用关于"三色猫"的文本,则输出以下结果:

```
dog - 0.28960436044424276
cat - 0.7103956395557574
```

其实,我们也可以直接使用 getBestCategory 方法来得到最好的分类。该方法使用一组输出结果,然后返回一个字符串。getAllResults 方法将所有的结果以字符串的形式返回。以下是这两个方法的示例:

```
System.out.println(categorizer.getBestCategory(outcomes));
System.out.println(categorizer.getAllResults(outcomes));
```

对应的输出为:

```
cat
dog[0.2896]   cat[0.7104]
```

6.4.2 Stanford API 的使用

Stanford API 支持多种分类器。本节中,我们将会演示如何使用 ColumnDataClassifier 类做一般分类,以及利用 StanfordCoreNLP 管道进行情感分析。Stanford API 支持的分类器有时不太容易使用。通过使用 ColumnDataClassifier 类,我们会演示如何分类盒子的尺寸。通过流水线的使用,我们会展示如何判断短语中包含的正面或负面情绪。分类器可以从以下地址下载:http://www.nlp.stanford.edu/wiki/Software/Classifier。

6.4.2.1 使用 ColumnDataClassifier 进行分类

该分类器使用多个数值描述数据。本小节中,我们通过使用训练文件来创建一个分类器,然后使用测试文件对该分类器的性能进行评估。ColumnDataClassifier 类通过使用属性文件来配置分类器的创建过程。

根据盒子的长、宽、高,我们要创建一个分类器用于盒子的分类。其类别为:小、中、大。将盒子特征的长、宽、高作为其特征,以浮点数描述。

属性文件指定了相关参数信息,并且提供了有关训练文件和测试文件的数据。许多属性都可以在该文件中指定。接下来,我们只使用其中一些比较相关的属性。

首先将随后要使用的属性文件存为 box.prop。第一组属性是关于训练文件和测试文件中特征的个数。由于要用到三个数值,所以要指定三个 realValued 列。trainFile 和 testFile 属性指定的是对应文件的存放位置和名称:

```
useClassFeature=true
1.realValued=true
2.realValued=true
3.realValued=true
trainFile=.box.train
testFile=.box.test
```

训练文件和测试文件使用的是相同的格式。每行包含类别名称以及由制表符分隔的数值。box.train 训练文件和 box.test 测试文件分别包含 60 个数据和 30 个数据。下面所示的是 box.train 文件的第一行,其类别是"小",高度、宽度、长度分别是 2.34、1.60、1.50:

```
small   2.34   1.60   1.50
```

创建分类器的代码如下所示。将属性文件作为构造器的参数来创建 ColumnDataClassifier 类的一个实例。makeClassifier 方法返回一个 Classifier 接口的实例。该接口支持三种方法,接下来我展示其中两种方法。readTrainingExamples 方法的作用是从训练文件中读取训练数据:

```
ColumnDataClassifier cdc =
    new ColumnDataClassifier("box.prop");
Classifier<String, String> classifier =
    cdc.makeClassifier(cdc.readTrainingExamples("box.train"));
```

这段代码执行后,会输出大量信息。这里我们只选部分信息进行讨论。输出结果的第一部分是属性文件的信息:

```
3.realValued = true
testFile = .box.test
...
trainFile = .box.train
```

接下来的部分显示数据集的个数和特征的信息:

```
Reading dataset from box.train ... done [0.1s, 60 items].
numDatums: 60
numLabels: 3 [small, medium, large]
...
AVEIMPROVE       The average improvement / current value
EVALSCORE        The last available eval score
Iter ## evals ## <SCALING> [LINESEARCH] VALUE TIME |GNORM| {RELNORM}
AVEIMPROVE EVALSCORE
```

然后分类器通过迭代数据完成创建过程:

```
Iter 1 evals 1 <D> [113M 3.107E-4] 5.985E1 0.00s |3.829E1| {1.959E-1}
0.000E0 -
Iter 2 evals 5 <D> [M 1.000E0] 5.949E1 0.01s |1.862E1| {9.525E-2} 3.058E-
3 -
Iter 3 evals 6 <D> [M 1.000E0] 5.923E1 0.01s |1.741E1| {8.904E-2} 3.485E-
3 -
...
Iter 21 evals 24 <D> [1M 2.850E-1] 3.306E1 0.02s |4.149E-1| {2.122E-3}
1.775E-4 -
Iter 22 evals 26 <D> [M 1.000E0] 3.306E1 0.02s
QNMinimizer terminated due to average improvement: | newest_val -
previous_val | / |newestVal| < TOL
Total time spent in optimization: 0.07s
```

此时,分类器已经创建完毕可以使用了。接下来,我们通过测试文件来验证该分类器。首先,使用 ObjectBank 类的 getLineIterator 方法获取文本文件的一行记录。该类可以将读入的数据转换成标准的格式。getLineIterator 方法每次可以返回一行适用于分类器格式的记录。该过程的循环操作如下:

```
for (String line :
        ObjectBank.getLineIterator("box.test", "utf-8")) {
    ...
}
```

在每个 for-each 语句中,一个 Datum 实例是由对应行记录创建的,然后它的 classOf 方法用来返回预测的类别,如下面代码所示。Datum 接口支持包含特征的对象,当其被用作 classOf 方法的参数时,分类器返回预测的类别:

```
Datum<String, String> datum = cdc.makeDatumFromLine(line);
System.out.println("Datum: {"
    + line + "]\tPredicted Category: "
    + classifier.classOf(datum));
```

这段代码执行后,如下所示,测试文件的每行记录经过处理,然后输出预测的类别。这里只展示最初两行和最后两行的记录。分类器准确地将测试数据进行了分类:

```
Datum: {small   1.33    3.50    5.43]   Predicted Category: medium
Datum: {small   1.18    1.73    3.14]   Predicted Category: small
...
Datum: {large   6.01    9.35    16.64]  Predicted Category: large
Datum: {large   6.76    9.66    15.44]  Predicted Category: large
```

如果只测试单个记录,我们可以使用 makeDatumFromStrings 方法创建一个 Datum 实例。在下面的代码中,创建一个一维的字符串数组,其中每个元素是盒子的相关数据,第一个元素为空,指代的是类别。然后将 Datum 实例作为 classOf 方法的参数来预测类别:

```
String sample[] = {"", "6.90", "9.8", "15.69"};
Datum<String, String> datum =
    cdc.makeDatumFromStrings(sample);
System.out.println("Category: " + classifier.classOf(datum));
```

这段代码的结果如下,其正确地预测出了盒子的类别:

```
Category: large
```

6.4.2.2 使用 Stanford 框架进行情感分析

本节将会阐述如何使用 Stanford API 进行情感分析。我们会使用 StanfordCoreNLP 框架在不同的文本上进行情感分析。

在这里,我们使用三段不同的文本,其中 review 字符串是一段来自"烂番茄"(Rotten Tomatoes,http://www.rottentomatoes.com/m/forrest_gump/)关于电影《阿甘正传》的影评:

```
String review = "An overly sentimental film with a somewhat "
    + "problematic message, but its sweetness and charm "
    + "are occasionally enough to approximate true depth "
    + "and grace. ";

String sam = "Sam was an odd sort of fellow. Not prone "
    + "to angry and not prone to merriment. Overall, "
    + "an odd fellow.";

String mary = "Mary thought that custard pie was the "
```

```
            + "best pie in the world. However, she loathed "
            + "chocolate pie.";
```

执行情感分析,我们需要使用一个如下所示的情感 annotator,同时也需要使用 tokenize、ssplit 和 parse 注解。其中,parse 注解为文本提供了更加结构化的信息,这些将在第 7 章中讨论:

```
Properties props = new Properties();
props.put("annotators", "tokenize, ssplit, parse, sentiment");
StanfordCoreNLP pipeline = new StanfordCoreNLP(props);
```

该文本用来创建一个 Annotation 实例,之后可用作 annotate 方法的参数来实现相应操作,如下所示:

```
Annotation annotation = new Annotation(review);
pipeline.annotate(annotation);
```

下面的数组存放的是不同情感的概率:

```
String[] sentimentText = {"Very Negative", "Negative",
    "Neutral", "Positive", "Very Positive"};
```

Annotation 类的 get 方法返回一个实现 CoreMap 接口的对象。这个示例中,这些对象指代的是将输入文本划分成句子的结果(参见以下代码)。对于每个句子,可以获取一个 Tree 对象的实例,该实例代表的是一个包含对于情感文本解析的树结构。getPredictedClass 方法将索引返回到 sentimentText 数组中用来指定测试的情感:

```
for (CoreMap sentence : annotation.get(
        CoreAnnotations.SentencesAnnotation.class)) {
    Tree tree = sentence.get(
        SentimentCoreAnnotations.AnnotatedTree.class);
    int score = RNNCoreAnnotations.getPredictedClass(tree);
    System.out.println(sentimentText[score]);
}
```

当使用 review 字符串执行代码后,可以得到以下结果:

Positive

sam 字符串文本包含三个句子,每个句子对应的结果如下:

Neutral

Negative

Neutral

mary 字符串文本包含两个句子,每个句子对应结果如下:

Positive

Neutral

6.4.3 使用 LingPipe 进行文本分类

本节中,我们使用 LingPipe 来演示一系列分类任务,包括一般的文本分类、情感分析以及语言识别。以下是将要涉及的分类主题:

- 使用 Classified 类训练文本
- 使用其他训练类别训练模型
- 如何使用 LingPipe 分类文本
- 使用 LingPipe 进行情感分析
- 识别所使用的语言

本节中描述的几个任务会用到如下声明。LingPipe 需要用到不同类别的训练数据,categories 数组包含由 LingPipe 打包的类别名称:

```
String[] categories = {"soc.religion.christian",
    "talk.religion.misc","alt.atheism","misc.forsale"};
```

DynamicLMClassifier 类用来完成实际的分类任务,它是通过包含类别名称的 categories 数组创建的。nGramSize 值指定的是分类模型的序列中相邻成员的数目:

```
int nGramSize = 6;
DynamicLMClassifier<NGramProcessLM> classifier =
    DynamicLMClassifier.createNGramProcess(
        categories, nGramSize);
```

6.4.3.1 使用分类类别训练文本

使用 LingPipe 进行一般的文本分类,首先通过训练集文件训练 DynamicLMClassifier 类,然后通过该类完成实际的分类任务。LingPipe 在其目录(demos/data/fourNewsGroups/4news-train)下提供了一些训练集。我们通过这些训练集来演示训练过程。以下示例是一个训练过程的简化版,完整版可从以下地址获取:http://alias-i.com/lingpipe/demos/tutorial/classify/read-me.html。

首先声明训练目录:

```
String directory = ".../demos";
File trainingDirectory = new File(directory
    + "/data/fourNewsGroups/4news-train");
```

训练目录中有 4 个子目录,它们的名称罗列在 categories 数组中。每个子目录包含一连串由数字命名的文件。这些文件含有用来处理目录和名称的"新闻组"(newsgroup)数据（http://qwone.com/~jason/20Newsgroups/）。

训练模型的过程要用到每个文件和类别以及 DynamicLMClassifier 类的 handle 方法。该方法利用文件为每个类别创建一个训练实例，然后通过实例扩充模型。这个过程会涉及 for 嵌套循环。

外层的 for 循环使用目录名称创建一个 File 对象，然后再调用它的 list 方法返回目录中的文件列表。每个文件的名称存放在内层 for 循环的 trainingFiles 数组中：

```java
for (int i = 0; i < categories.length; ++i) {
    File classDir =
        new File(trainingDirectory, categories[i]);
    String[] trainingFiles = classDir.list();
    // Inner for-loop
}
```

如下所示的内层 for 循环打开每个文件，并读取其中的文本。Classification 类将分类表示成特定的类别，它是和文本一起使用来创建 Classified 实例。DynamicLMClassifier 类的 handle 方法根据新的信息更新模型：

```java
for (int j = 0; j < trainingFiles.length; ++j) {
    try {
        File file = new File(classDir, trainingFiles[j]);
        String text = Files.readFromFile(file, "ISO-8859-1");
        Classification classification =
            new Classification(categories[i]);
        Classified<CharSequence> classified =
            new Classified<>(text, classification);
        classifier.handle(classified);
    } catch (IOException ex) {
        // Handle exceptions
    }
}
```

> 读者可使用 java.io.File 中的 com.aliasi.util.Files 类，否则无法使用 readFromFile 方法。

如下所示，我们可将分类器序列化以便后续使用。AbstractExternalizable 类是一个支持对象序列化的工具类。它有一个静态 compileTo 方法，该方法可接受一个 Compilable 实例和一个 File 对象。如下所示，它可以将对象序列化成文件：

```
try {
    AbstractExternalizable.compileTo( (Compilable) classifier,
        new File("classifier.model"));
} catch (IOException ex) {
    // Handle exceptions
}
```

分类器的加载将会在本章 6.4.3.3 节中说明。

6.4.3.2 使用其他的训练类别

在 http://qwone.com/~jason/20Newsgroups/ 中可以找到其他的 "新闻组" 数据。这些数据可以用来训练以下表格中的分类器。尽管只有 20 个类别，但它们都是非常有用的训练模型。三组可下载的数据中，其中一些是已经排序过的，其他组中去掉了重复的数据：

新闻组	
comp.graphics	sci.crypt
comp.os.ms-windows.misc	sci.electronics
comp.sys.ibm.pc.hardware	sci.med
comp.sys.mac.hardware	sci.space
comp.windows.x	misc.forsale
rec.autos	talk.politics.misc
rec.motorcycles	talk.politics.guns
rec.sport.baseball	talk.politics.mideast
rec.sport.hockey	talk.religion.misc
alt.atheism	

6.4.3.3 LingPipe 的文本分类

为了分类文本，我们将用到 DynamicLMClassifier 类的 classify 方法。下面我们会通过两段文本来演示其用法：

- forSale：第一段文本来自 http://www.homes.com/for-sale/ 的第一个完整句。
- martinLuther：第二段文本来自 https://en.wikipedia.org/wiki/Martin_Luther 的第二段中的第一句。

文本声明如下：

```
String forSale =
    "Finding a home for sale has never been "
```

```
        + "easier. With Homes.com, you can search new "
        + "homes, foreclosures, multi-family homes, "
        + "as well as condos and townhouses for sale. "
        + "You can even search our real estate agent "
        + "directory to work with a professional "
        + "Realtor and find your perfect home.";
String martinLuther =
    "Luther taught that salvation and subsequently "
        + "eternity in heaven is not earned by good deeds "
        + "but is received only as a free gift of God's "
        + "grace through faith in Jesus Christ as redeemer "
        + "from sin and subsequently eternity in Hell.";
```

为使用前面小节中序列化后的分类器，这里使用 AbstractExternalizable 类的 readObject 方法，并且使用 LMClassifier 类而不是 DynamicLMClassifier 类。它们都支持 classify 方法，区别在于 DynamicLMClassifier 类不可以被序列化：

```
LMClassifier classifier = null;
try {
    classifier = (LMClassifier)
        AbstractExternalizable.readObject(
            new File("classifier.model"));
} catch (IOException | ClassNotFoundException ex) {
    // Handle exceptions
}
```

在下面的代码段中，我们应用 LMClassifier 类的 classify 方法，返回的是一个可用于判断最优匹配的 JointClassification 实例：

```
JointClassification classification =
    classifier.classify(text);
System.out.println("Text: " + text);
String bestCategory = classification.bestCategory();
System.out.println("Best Category: " + bestCategory);
```

对于下面的 forSale 文本，得到如下结果：

```
Text: Finding a home for sale has never been easier. With Homes.com,
you can search new homes, foreclosures, multi-family homes, as well as
condos and townhouses for sale. You can even search our real estate agent
directory to work with a professional Realtor and find your perfect home.
Best Category: misc.forsale
```

对于 martinLuther 文本，得到如下结果：

```
Text: Luther taught that salvation and subsequently eternity in heaven
is not earned by good deeds but is received only as a free gift of God's
grace through faith in Jesus Christ as redeemer from sin and subsequently
eternity in Hell.
Best Category: soc.religion.christian
```

这些文本都被正确地分类了。

6.4.3.4 使用 LingPipe 进行情感分析

情感分析的处理过程和一般的文本分类过程非常相似。有一处不同的是，情感分析使用的是两种类别："正面"和"负面"。

我们需要使用数据文件训练模型。我们将使用由 http://alias-i.com/lingpipe/demos/tutorial/sentiment/read-me.html 提供的情感分析的简化版本，其使用的情感数据是为电影建立的（http://www.cs.cornell.edu/people/pabo/movie-review-data/review_polarity.tar.gz）。这个数据是来自互联网电影数据库（IMDb）中的 1000 条正面和 1000 条负面的影评。

这些影评需要下载和解压。解压后，会出现一个 txt_sentoken 目录，其包含两个子目录，分别是：neg 和 pos。这两个子目录中都包含影评数据。尽管这些影评数据的一部分就能评测创建的模型，但为了简化解释说明，我们将使用全部的数据用于评测。

首先，重新初始化在 6.4.3 节中声明的变量。将 categories 声明为包含两个元素的数组用于存储两个类别。赋予 classifier 变量一个新的 DynamicLMClassifier 实例，参数分别为新的类别数组变量和值为 8 的 nGramSize 变量。

```
categories = new String[2];
categories[0] = "neg";
categories[1] = "pos";
nGramSize = 8;
classifier = DynamicLMClassifier.createNGramProcess(
    categories, nGramSize);
```

与之前的做法相同，我们基于训练文件中的内容创建一系列实例。我们不详细解释下面的代码，因为它和 6.4.3.1 节中的代码非常类似，主要区别在于，这里只有两种类别需要处理：

```
String directory = "...";
File trainingDirectory = new File(directory, "txt_sentoken");
for (int i = 0; i < categories.length; ++i) {
    Classification classification =
        new Classification(categories[i]);
    File file = new File(trainingDirectory, categories[i]);
    File[] trainingFiles = file.listFiles();
    for (int j = 0; j < trainingFiles.length; ++j) {
        try {
```

```
                String review = Files.readFromFile(
                    trainingFiles[j], "ISO-8859-1");
                Classified<CharSequence> classified =
                    new Classified<>(review, classification);
                classifier.handle(classified);
            } catch (IOException ex) {
                ex.printStackTrace();
            }
        }
    }
```

此时模型已经可以使用了,首先使用的影评来自《阿甘正传》:

```
String review = "An overly sentimental film with a somewhat "
    + "problematic message, but its sweetness and charm "
    + "are occasionally enough to approximate true depth "
    + "and grace. ";
```

我们使用 classify 方法进行实际的分类工作。它返回的是 Classification 实例,其 bestCategory 方法返回的是最优类别,如下所示:

```
Classification classification = classifier.classify(review);
String bestCategory = classification.bestCategory();
System.out.println("Best Category: " + bestCategory);
```

代码执行后,得到如下结果:

```
Best Category: pos
```

该方法同样适用于其他类别的文本。

6.4.3.5 使用 LingPipe 进行语言识别

LingPipe 在其目录 demos/models 下有一个模型 langid-leipzig.classifier,它是通过多种语言训练得到的。下面表格中列出的是它支持的语言。该模型使用的训练数据来自 Leipzig 语料库(http://corpora.uni-leipzig.de/)。另外一个实用的工具可以 https://code.google.com/p/language-detection/ 找到。

语言	缩写	语言	缩写
Catalan	cat	Italian	it
Danish	dk	Japanese	jp
English	en	Korean	kr
Estonian	ee	Norwegian	no
Finnish	fi	Sorbian	sorb
French	fr	Swedish	se
German	de	Turkish	tr

为了使用该模型，我们调用 6.4.3.3 节中的相同代码。首先使用来自《阿甘正传》的相同影评：

```
String text = "An overly sentimental film with a somewhat "
    + "problematic message, but its sweetness and charm "
    + "are occasionally enough to approximate true depth "
    + "and grace. ";
System.out.println("Text: " + text);
```

随后使用 langid-leipzig.classifier 文件创建一个 LMClassifier 实例：

```
LMClassifier classifier = null;
try {
    classifier = (LMClassifier)
        AbstractExternalizable.readObject(
            new File(".../langid-leipzig.classifier"));
} catch (IOException | ClassNotFoundException ex) {
    // Handle exceptions
}
```

使用 classifier 方法后调用 bestCategory 方法获取最佳匹配的语言类别，如下所示：

```
Classification classification = classifier.classify(text);
String bestCategory = classification.bestCategory();
System.out.println("Best Language: " + bestCategory);
```

如果选择"英语"的文本，结果如下：

Text: An overly sentimental film with a somewhat problematic message, but its sweetness and charm are occasionally enough to approximate true depth and grace.
Best Language: en

下面的代码示例使用的是瑞典维基百科（http://sv.wikipedia.org/wiki/Svenska）中的第一个句子：

```
text = "Svenska är ett östnordiskt språk som talas av cirka "
    + "tio miljoner personer[1], främst i Finland "
    + "och Sverige.";
```

如下的结果正确地识别出了其语言类别：

Text: Svenska är ett östnordiskt språk som talas av cirka tio miljoner personer[1], främst i Finland och Sverige.
Best Language: se

其训练过程和之前介绍的训练 LingPipe 模型的过程是一样的。进行语言识别时，需要注意若文本包含多种语言，识别过程会变得复杂。

6.5 本章小结

在本章中，我们讨论了文本分类涉及的问题，同时也使用了不同的方法实现了该过程。文本分类对很多应用都是非常有用的，比如：检测垃圾邮件，鉴别一篇文档的可能作者，识别性别以及识别语言。

同时我们也展示了执行情感分析的过程。情感分析主要关注的是一段文本中的情感是正面的还是负面的。除此之外，也可扩展到其他的情感属性上。

我们使用的大多数方法首先需要基于训练数据创建一个模型。通常情况下，该模型需要通过一组测试数据进行验证。一旦模型创建完毕，它的使用过程通常很简单。

在下一章中，我们将会考察文本解析过程以及它在关系提取中的作用。

CHAPTER 7

第 7 章

关 系 提 取

通过对文本单元的解析可以建立解析树,解析机器语言十分简单(毕竟就是机器的语言),然而写代码是很困难的,更不必说自然语言的解析了。自然语言具有歧义性,歧义性使语言难以学习,但具有极大的灵活性和表达力。在这里,我们对解析机器语言不感兴趣,感兴趣的是自然语言的解析。

解析树是表示句子语法结构的层次化数据结构。通常,这是一个带有根节点的树图,后续内容将对其说明,我们将使用解析树帮助识别其中实体之间的关系。

解析可用于多种任务,包括:

- 机器翻译
- 语音合成
- 语音识别
- 语法检查
- 信息提取

共指消解(coreference resolution)是指在文本中的两个或两个以上表达式指向同一个人或物的情况。例如,在以下这句话中:

"Ted went to the party where he made an utter fool of himself."

"Ted""he"和"himself"都指的是"Ted",这对于正确解读文本和确定文本的相关重要性程度是非常重要的。下文将说明如何使用 Stanford 的 API 解决这个问题。

从文本中提取关系和有用的信息是 NLP 中一项重要的任务。实体之间（比如一个句子的主语和它的宾语、其他实体或它的行为之间）可能存在各种关系。我们可以直接使用所得结果或者进行格式化以便更好地利用它们来完成下游任务。

这一章介绍解析文本的过程及解析树的使用，包括关系提取、关系类型研究、关系提取应用和 NLP 的 API 的使用。

7.1 关系类型

事物之间的关系有多种，下表中列出了一些类别及其例子。更多详细的内容可参考 Freebase 网站（https://www.freebase.com/），这一数据库中按照人物、地点和事物等分门别类。另外还可以参考一下 WordNet 辞典（http://wordnet.princeton.edu/）。

关系	例子
个人的	father-of、sister-of、girlfriend-of
组织的	subsidiary-of、subcommittee-of
空间的	near-to、northeast-of、under
物质的	part-of、composed-of
相互作用	bonds-with、associates-with、reacts-with

命名实体识别（NER）是第 4 章介绍过的较低水平的 NLP 分类任务。然而，许多应用不仅需要如此，还希望识别不同的关系类型。当 NER 识别出一个实体后，如果可以判断这是一个人，那么对于进一步的关系提取十分有利。

一旦确定了这些实体，就可以将链接创建到它们包含的文档中或用作索引。对于问答应用，回答中通常用到命名实体。当确定了文本的情感特征时，这一情感也要归到一些实体上。

比如以下输出：

```
He was the last person to see Fred.
```

使用第 4 章介绍的 OpenNLP NER 进行处理，结果如下：

```
Span: [7..9) person
Entity: Fred
```

使用 OpenNLP 解析器，得到句子的更多信息：

```
(TOP (S (NP (PRP He)) (VP (VBD was) (NP (NP (DT the) (JJ last) (NN
person)) (SBAR (S (VP (TO to) (VP (VB see)))))) (. Fred.)))
```

再看下面这句话：

```
The cow jumped over the moon.
```

解析结果如下：

```
(TOP (S (NP (DT The) (NN cow)) (VP (VBD jumped) (PP (IN over) (NP (DT
the) (NN moon))))))
```

解析分为两种。

- 依赖型：关注于单词之间的关系
- 解析结构型：处理词组及其递归结构

依赖型使用主语、限定词、介词等提取关系。解析方法包括 shift-reduce、spanning tree 和 cascaded chunking。这里不讨论它们之间的差异，而是要研究它们的使用方法和输出结果。

7.2 理解解析树

解析树代表文本元素间层次化的关系。依赖型树展示了句子语法元素之间的关系。看下面这句话：

```
The cow jumped over the moon.
```

这句话的解析树如下，这是由 7.5.2.1 节中使用的方法生成的：

```
(ROOT
  (S
    (NP (DT The) (NN cow))
(VP (VBD jumped)
  (PP (IN over)
    (NP (DT the) (NN moon))))
(. .)))
```

这个句子可以描述为下图中的图像，这是由 http://nlpviz.bpodgursky.com/home 中的应用程序生成的。另外这个图还可以使用斯坦福大学支持的 GrammarScope 编辑器（http://grammarscope.sourceforge.net/），它使用基于 Swing 的图形用户界面，可以生成解析树、语法结构、依赖关系和文本的语义图。

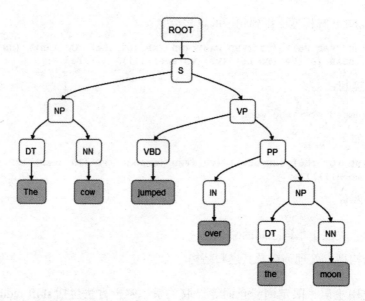

然而，解析句子的方法不止一种，解析不是一件容易的事情，尤其是解析可能存在许多歧义的一大段文本。下面列出了用其他方法对于之前的示例句子进行解析得到的依赖型树。使用OpenNLP生成的解析树，在本章7.5.1节中还有介绍：

```
(TOP (S (NP (DT The) (NN cow)) (VP (VBD jumped) (PP (IN over) (NP (DT
the) (NN moon))))))
(TOP (S (NP (DT The) (NN cow)) (VP (VP (VBD jumped) (PRT (RP over))) (NP
(DT the) (NN moon)))))
(TOP (S (NP (DT The) (NNS cow)) (VP (VBD jumped) (PP (IN over) (NP (DT
the) (NN moon))))))
```

对同一个句子的解析，以上所得结果均略有差异，最可靠的是第一个。

7.3 关系提取的应用

提取到的关系可以用于以下几种目的：

- 建立知识库
- 创建目录
- 产品搜索
- 专利分析
- 股票研究
- 情报分析

下图所示是一个维基百科的信息框，当进入 Oklahoma 词条时，可以在这个信息框中看到列出的各种关系类型，如官方语言、行政中心及该区域的具体情况。

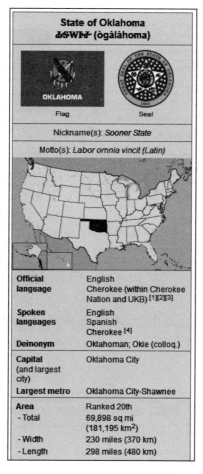

有许多用于关系、信息提取的数据库是根据维基百科建立的，如下例所示。

- Resource Description Framework（RDF）：采用三元组，如 Yosemite-location-California，其中，location 表示其关联，其网站为 http://www.w3.org/RDF/。
- DBPedia：存储有十亿以上的三元组，是一个根据维基百科建立的知识库，其网站为 http://dbpedia.org/About。

另一个简单有趣的例子如下图所示，当用谷歌搜索"水星"（planet mercury）时，我们不仅获得了一些查询结果的链接，还看到了页面右侧显示的水星的关系信息和图片：

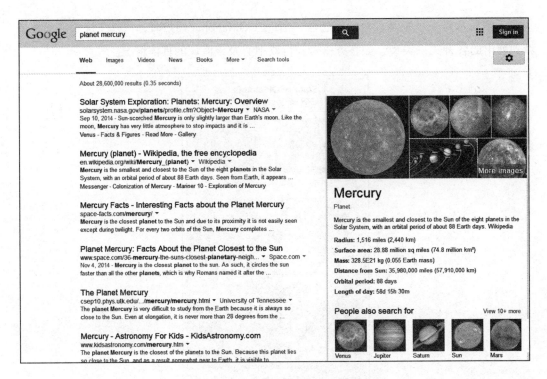

信息提取也用于创建 Web 索引，这些网站的索引方便用户浏览这个网站。下面的截图是美国人口普查局网站（http://www.census.gov/main/www/a2z）的 Web 索引的例子：

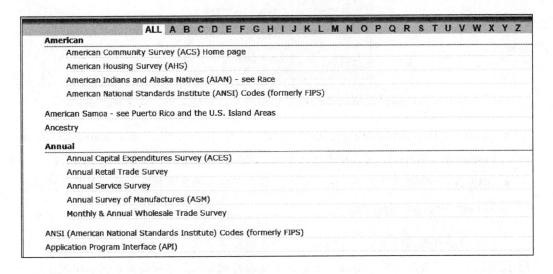

7.4 关系提取

关系提取有许多可行的方法，可以分类如下：

- 手工模式
- 监督方法
- 半监督或无监督方法
    - 自提升法
    - 远程监督法
    - 无监督法

当我们没有训练数据时，比如一个新的业务领域或完全新类型的项目，此时应该用手工模型。这种模型通常需要一定的规则，比如：含有"演员"而不含有"电影"广告"的文本该属于"剧本"。

然而，这种方法耗时耗力，需要随实际文本而调整。

如果有一些不错的训练数据，可以尝试使用朴素贝叶斯方法；如果有更多的数据，SVM、正则化的逻辑回归和随机森林等方法都可以使用。

虽然也有必要了解这些方法的原理，但是此处重点说明它们的使用方法。

7.5 使用 NLP API

我们将使用 OpenNLP 和 Stanford 的 API 演示文本解析和关系提取的方法。LingPipe 可用于解析生物医药专业文献，本书不做介绍，详情可参考 http://alias-i.com/lingpipe-3.9.3/demos/tutorial/medline/read-me.html。

7.5.1 OpenNLP 的使用

ParserTool 类可以简单实现文本解析。静态的 parseLine 方法接受三个参数，返回一个 Parser 实例。这三个参数是：

- 待解析的字符串
- Parser 实例

- 指定返回的解析内容数量的一个整数

Parser 实例存储所有解析的元素,按其可能性的大小顺序返回。创建一个 Parser 实例需使用 ParserFactory 类的 create 方法,该方法使用借助 en-parser-chunking.bin 文件创建的 ParserModel 实例。

代码如下所示,用 try-with-resources 语句块创建模型文件的输入流,创建 ParserModel 实例以及一个 Parser 实例:

```
String fileLocation = getModelDir() + 
    "/en-parser-chunking.bin";
try (InputStream modelInputStream = 
            new FileInputStream(fileLocation);) {
    ParserModel model = new ParserModel(modelInputStream);
    Parser parser = ParserFactory.create(model);
    ...
} catch (IOException ex) {
    // Handle exceptions
}
```

解析过程的代码十分简单,如下所示,调用 parseLine 方法,第三个参数是 3,表示将返回前三个解析内容:

```
String sentence = "The cow jumped over the moon";
Parse parses[] = ParserTool.parseLine(sentence, parser, 3);
```

接下来,列出解析内容及其可能性大小,如下所示:

```
for(Parse parse : parses) {
    parse.show();
    System.out.println("Probability: " + parse.getProb());
}
```

所得结果如下:

```
(TOP (S (NP (DT The) (NN cow)) (VP (VBD jumped) (PP (IN over) (NP (DT
the) (NN moon))))))
Probability: -1.043506016751117
(TOP (S (NP (DT The) (NN cow)) (VP (VP (VBD jumped) (PRT (RP over))) (NP
(DT the) (NN moon)))))
Probability: -4.248553665013661
(TOP (S (NP (DT The) (NNS cow)) (VP (VBD jumped) (PP (IN over) (NP (DT
the) (NN moon))))))
Probability: -4.761071294573854
```

每个解析的顺序和标注均有所不同。我们将第一个解析格式化以便更易于理解,如下:

```
(TOP
    (S
        (NP
            (DT The)
            (NN cow)
        )
        (VP
            (VBD jumped)
            (PP
                (IN over)
                (NP
                    (DT the)
                    (NN moon)
                )
            )
        )
    )
)
```

showCodeTree 方法可以用于显示父子元素的关系：

```
parse.showCodeTree();
```

第一个解析的输出如下，每行第一部分是括号内元素的级别，随后是由 -> 隔开的两个哈希值，最后是标注。哈希值分别对应该元素及其父元素。比如第三行为专有名词"The"，其父元素为名词短语"The cow"：

```
[0] S -929208263 -> -929208263 TOP The cow jumped over the moon
[0.0] NP -929237012 -> -929208263 S The cow
[0.0.0] DT -929242488 -> -929237012 NP The
[0.0.0.0] TK -929242488 -> -929242488 DT The
[0.0.1] NN -929034400 -> -929237012 NP cow
[0.0.1.0] TK -929034400 -> -929034400 NN cow
[0.1] VP -928803039 -> -929208263 S jumped over the moon
[0.1.0] VBD -928822205 -> -928803039 VP jumped
[0.1.0.0] TK -928822205 -> -928822205 VBD jumped
[0.1.1] PP -928448468 -> -928803039 VP over the moon
[0.1.1.0] IN -928460789 -> -928448468 PP over
[0.1.1.0.0] TK -928460789 -> -928460789 IN over
[0.1.1.1] NP -928195203 -> -928448468 PP the moon
[0.1.1.1.0] DT -928202048 -> -928195203 NP the
[0.1.1.1.0.0] TK -928202048 -> -928202048 DT the
[0.1.1.1.1] NN -927992591 -> -928195203 NP moon
[0.1.1.1.1.0] TK -927992591 -> -927992591 NN moon
```

通过 getChildren 方法也可得到解析结果的元素,该方法返回一个 Parser 对象数组,每个对象是解析结果的一个元素。使用 Parse 方法可以得到每个元素的文本、标注和标签,如下所示:

```
Parse children[] = parse.getChildren();
for (Parse parseElement : children) {
    System.out.println(parseElement.getText());
    System.out.println(parseElement.getType());
    Parse tags[] = parseElement.getTagNodes();
    System.out.println("Tags");
    for (Parse tag : tags) {
        System.out.println("[" + tag + "]"
            + " type: " + tag.getType()
            + "  Probability: " + tag.getProb()
            + "  Label: " + tag.getLabel());
    }
}
```

所得结果如下:

```
The cow jumped over the moon
S
Tags
[The] type: DT  Probability: 0.9380626549164167  Label: null
[cow] type: NN  Probability: 0.9574993337971017  Label: null
[jumped] type: VBD  Probability: 0.9652983971550483  Label: S-VP
[over] type: IN  Probability: 0.7990638213315913  Label: S-PP
[the] type: DT  Probability: 0.9848023215770413  Label: null
[moon] type: NN  Probability: 0.9942338356992393  Label: null
```

7.5.2 使用 Stanford API

Stanford NLP 的 API 有多个文本解析的方法。首先我们演示一个一般目的的解析器(LexicalizedParser 类)。然后,介绍如何使用 TreePrint 类列出解析的结果。最后,演示用 GrammaticalStructure 类决定单词间的依赖关系。

7.5.2.1 LexicalizedParser 类的使用

LexicalizedParser 类是一个词汇化的 PCFG 解析器,可以使用多种不同的模型。使用 apply 方法可以建立一棵解析树,其参数是一个 CoreLabel 对象的 List 实例。

首先,使用 englishPCFG.ser.gz 的模型初始化解析器:

```
String parserModel = ".../models/lexparser/englishPCFG.ser.gz";
LexicalizedParser lexicalizedParser =
    LexicalizedParser.loadModel(parserModel);
```

用 Sentence 类的 toCoreLabelList 方法创建 CoreLabel 对象的 List 实例。其中 CoreLabel 对象包括一个单词及其他信息，并没有标注或标签，实际上其中的单词已被分词处理了。

```
String[] senetenceArray = {"The", "cow", "jumped", "over",
    "the", "moon", "."};
List<CoreLabel> words =
    Sentence.toCoreLabelList(senetenceArray);
```

此时调用 apply 方法：

```
Tree parseTree = lexicalizedParser.apply(words);
```

使用 pennPrint 方法可以很方便地列出解析结果，与 Penn TreeBank 所用方法相同（http://www.sfs.uni-tuebingen.de/~dm/07/autumn/795.10/ptb-annotationguide/root.html）：

```
parseTree.pennPrint();
```

所得结果如下：

```
(ROOT
  (S
    (NP (DT The) (NN cow))
    (VP (VBD jumped)
      (PP (IN over)
        (NP (DT the) (NN moon))))
    (. .)))
```

Tree 类提供了很多处理解析树的方法。

7.5.2.2 TreePrint 方法的使用

TreePrint 类提供很简单的方法列出解析树。创建一个实例，使用描述所需格式的字符串为参数，使用静态的 outputTreeFormats 变量可得到有效的输出格式的数字，如下表所示：

Tree 格式字符串		
penn	dependencies	collocations
oneline	typedDependencies	semanticGraph
rootSymbolOnly	typedDependenciesCollapsed	conllStyleDependencies
words	latexTree	conll2007
wordsAndTags	xmlTree	

斯坦福大学使用类型依赖性来描述句子内部存在的语法关系，详情可参考 Stanford Typed Dependencies Manual（http://nlp.stanford.edu/software/dependencies_manual.pdf）。

下列代码演示了 TreePrint 类的使用方式，通过 printTree 方法执行实际的输出操作。

这种情况下创建了 TreePrint 对象，带有类型依赖性"collapsed"。

```
TreePrint treePrint =
    new TreePrint("typedDependenciesCollapsed");
treePrint.printTree(parseTree);
```

输出结果如下，其中的数字表示它在句子中的位置：

```
det(cow-2, The-1)
nsubj(jumped-3, cow-2)
root(ROOT-0, jumped-3)
det(moon-6, the-5)
prep_over(jumped-3, moon-6)
```

使用字符串"penn"创建对象可得到以下结果：

```
(ROOT
  (S
    (NP (DT The) (NN cow))
    (VP (VBD jumped)
      (PP (IN over)
        (NP (DT the) (NN moon))))
    (. .)))
```

字符串"dependencies"生成了依赖关系的一个简单列表：

```
dep(cow-2,The-1)
dep(jumped-3,cow-2)
dep(null-0,jumped-3,root)
dep(jumped-3,over-4)
dep(moon-6,the-5)
dep(over-4,moon-6)
```

可以使用逗号合并格式，下面的代码将得到 penn 类型和 typedDependenciesCollapsed 格式的结果：

```
"penn,typedDependenciesCollapsed"
```

7.5.2.3 使用 GrammaticalStructure 类建立单词依赖关系

文本解析可以使用前一节中创建的 LexicalizedParser 对象，并结合 Treebank-

LanguagePack 接口。树图资料库（Treebank）是一个注释了语法和语义信息的文本语料库，提供了一个句子结构的信息，可以通过人工或半自动方式创建。最大的树图资料库是 Penn TreeBank（http://www.cis.upenn.edu/~treebank/）。

下面的例子说明了如何使用解析器格式化一个简单的字符串。分词器工厂用于创建一个分词器，还用到了 7.5.2.1 节中的 CoreLabel 类：

```
String sentence = "The cow jumped over the moon.";
TokenizerFactory<CoreLabel> tokenizerFactory =
    PTBTokenizer.factory(new CoreLabelTokenFactory(), "");
Tokenizer<CoreLabel> tokenizer =
    tokenizerFactory.getTokenizer(new StringReader(sentence));
List<CoreLabel> wordList = tokenizer.tokenize();
parseTree = lexicalizedParser.apply(wordList);
```

TreebankLanguagePack 接口指定了使用 Treebank 的方法。在下面的代码中，创建了一系列对象，最终创建一个 TypedDependency 实例，该实例用于获取有关句子元素的依赖关系信息。创建一个 GrammaticalStructureFactory 对象的实例，并用于创建一个 GrammaticalStructure 类实例：

```
TreebankLanguagePack tlp =
    lexicalizedParser.treebankLanguagePack;
GrammaticalStructureFactory gsf =
    tlp.grammaticalStructureFactory();
GrammaticalStructure gs =
    gsf.newGrammaticalStructure(parseTree);
List<TypedDependency> tdl = gs.typedDependenciesCCprocessed();
```

如下代码可以列出结果：

```
System.out.println(tdl);
```

结果如下：

```
[det(cow-2, The-1), nsubj(jumped-3, cow-2), root(ROOT-0, jumped-3),
det(moon-6, the-5), prep_over(jumped-3, moon-6)]
```

可以通过 gov、reln 和 dep 方法提取信息，分别可以得到核心词、关系和依赖的元素，说明如下：

```
for(TypedDependency dependency : tdl) {
    System.out.println("Governor Word: [" + dependency.gov()
        + "] Relation: [" + dependency.reln().getLongName()
        + "] Dependent Word: [" + dependency.dep() + "]");
}
```

结果如下：

```
Governor Word: [cow/NN] Relation: [determiner] Dependent Word: [The/DT]
Governor Word: [jumped/VBD] Relation: [nominal subject] Dependent Word:
[cow/NN]
Governor Word: [ROOT] Relation: [root] Dependent Word: [jumped/VBD]
Governor Word: [moon/NN] Relation: [determiner] Dependent Word: [the/DT]
Governor Word: [jumped/VBD] Relation: [prep_collapsed] Dependent Word:
[moon/NN]
```

据此可以对句子内的元素间关系有个大致了解。

7.5.3 判断共指消解的实体

共指消解是指文本中两个及以上的表达式指向同一个人或实体的情况。看下面的例子：

"He took his cash and she took her change and together they bought their lunch."

这句话中有多个共指关系，"his"指的是"He"，"her"指的是"she"，还有"they"指的是"He"和"she"。

语境照应词（endophora）是一个在其之前或之后的表达式的共指关系，可以被分为向前照应词（anaphor）和向后照应词（cataphor）。下面这个句子"Mary felt the earthquake. It shook the entire building."中，"It"是先行词（antecedent）"the earthquake"的向前照应词。而在这个句子"As she sat there, Mary felt the earthquake."中，"she"是后行词（postcedent）"Mary"的向后照应词。

通过斯坦福大学的 API 内 StanfordCoreNLP 类的 dcoref 注释可以解决共指消解，我们将以之前的句子为例进行说明。

我们首先建立流水线，使用 annotate 方法，如下所示：

```
String sentence = "He took his cash and she took her change "
    + "and together they bought their lunch.";
Properties props = new Properties();
props.put("annotators",
    "tokenize, ssplit, pos, lemma, ner, parse, dcoref");
StanfordCoreNLP pipeline = new StanfordCoreNLP(props);
Annotation annotation = new Annotation(sentence);
pipeline.annotate(annotation);
```

使用 Annotation 类的 get 方法，传入 CorefChainAnnotation.class 作为参数，可以得

到 CorefChain 对象的一个 Map 实例，其中存储了句子内找到的共指关系的信息：

```
Map<Integer, CorefChain> corefChainMap =
    annotation.get(CorefChainAnnotation.class);
```

CorefChain 对象的集合使用整数索引，如下所示对这些对象进行遍历，得到键集便可以列出每一个 CorefChain 对象了：

```
Set<Integer> set = corefChainMap.keySet();
Iterator<Integer> setIterator = set.iterator();
while(setIterator.hasNext()) {
    CorefChain corefChain =
        corefChainMap.get(setIterator.next());
    System.out.println("CorefChain: " + corefChain);
}
```

结果如下所示：

```
CorefChain: CHAIN1-["He" in sentence 1, "his" in sentence 1]
CorefChain: CHAIN2-["his cash" in sentence 1]
CorefChain: CHAIN4-["she" in sentence 1, "her" in sentence 1]
CorefChain: CHAIN5-["her change" in sentence 1]
CorefChain: CHAIN7-["they" in sentence 1, "their" in sentence 1]
CorefChain: CHAIN8-["their lunch" in sentence 1]
```

使用 CorefChain 和 CorefMention 类的方法可以得到更多详细的信息。后者包含了句子内找到的具体共指关系的信息。

在前面代码的 while 循环主体中加入下列代码，可以得到并列出所需信息，该类的 startIndex 和 endIndex 两个成员变量指的是句子中单词的位置。

```
System.out.print("ClusterId: " + corefChain.getChainID());
CorefMention mention = corefChain.getRepresentativeMention();
System.out.println(" CorefMention: " + mention
    + " Span: [" + mention.mentionSpan + "]");

List<CorefMention> mentionList =
    corefChain.getMentionsInTextualOrder();
Iterator<CorefMention> mentionIterator =
    mentionList.iterator();
while(mentionIterator.hasNext()) {
    CorefMention cfm = mentionIterator.next();
    System.out.println("\tMention: " + cfm
        + " Span: [" + mention.mentionSpan + "]");
    System.out.print("\tMention Mention Type: "
        + cfm.mentionType + " Gender: " + cfm.gender);
    System.out.println(" Start: " + cfm.startIndex
```

```
            + " End: " + cfm.endIndex);
}
System.out.println();
```

结果如下所示,在此只列出了第一个和最后一个:

```
CorefChain: CHAIN1-["He" in sentence 1, "his" in sentence 1]
ClusterId: 1 CorefMention: "He" in sentence 1 Span: [He]
  Mention: "He" in sentence 1 Span: [He]
  Mention Type: PRONOMINAL Gender: MALE Start: 1 End: 2
  Mention: "his" in sentence 1 Span: [He]
  Mention Type: PRONOMINAL Gender: MALE Start: 3 End: 4
…
CorefChain: CHAIN8-["their lunch" in sentence 1]
ClusterId: 8 CorefMention: "their lunch" in sentence 1 Span: [their
lunch]
  Mention: "their lunch" in sentence 1 Span: [their lunch]
  Mention Type: NOMINAL Gender: UNKNOWN Start: 14 End: 16
```

7.6 问答系统的关系提取

这一节中我们介绍可以用于问答机制的关系提取方法。需要回答的问题可能是这样:

- 美国的第 14 任总统是谁?
- 美国现任总统来自哪里?
- 奥巴马总统在任时间是什么时候?

回答这些类型的问题并不容易。我们将演示回答某些类型问题的一种方法,但我们将对过程的许多方面进行简化。既使有这些限制,我们会发现对问题的回答效果还不错。

这一过程包括以下几个步骤:

1)判断单词的依赖关系

2)识别问题的类型

3)提取相关成分

4)搜索答案

5)回答问题

我们将演示判断问题类型（who/what/when/where）的一般框架。然后，我们研究回答"who"类型问题的一些话题。

为了简化问题，我们将问题限制为关于美国总统的问题，这样可以使用一个简单的相关数据库来查询问题的答案。

7.6.1 判断单词依赖关系

将问题存储为一个简单的字符串：

```
String question =
    "Who is the 32nd president of the United States?";
```

按 7.5.2.3 节中所示使用 LexicalizedParser 类，为了方便复制代码如下：

```
String parserModel = ".../englishPCFG.ser.gz";
LexicalizedParser lexicalizedParser =
    LexicalizedParser.loadModel(parserModel);

TokenizerFactory<CoreLabel> tokenizerFactory =
    PTBTokenizer.factory(new CoreLabelTokenFactory(), "");
Tokenizer<CoreLabel> tokenizer =
    tokenizerFactory.getTokenizer(new StringReader(question));
List<CoreLabel> wordList = tokenizer.tokenize();
Tree parseTree = lexicalizedParser.apply(wordList);

TreebankLanguagePack tlp =
    lexicalizedParser.treebankLanguagePack();
GrammaticalStructureFactory gsf =
    tlp.grammaticalStructureFactory();
GrammaticalStructure gs =
    gsf.newGrammaticalStructure(parseTree);
List<TypedDependency> tdl = gs.typedDependenciesCCprocessed();
System.out.println(tdl);
for (TypedDependency dependency : tdl) {
    System.out.println("Governor Word: [" + dependency.gov()
        + "] Relation: [" + dependency.reln().getLongName()
        + "] Dependent Word: [" + dependency.dep() + "]");
}
```

执行后结果如下：

```
[root(ROOT-0, Who-1), cop(Who-1, is-2), det(president-5, the-3),
amod(president-5, 32nd-4), nsubj(Who-1, president-5), det(States-9, the-
7), nn(States-9, United-8), prep_of(president-5, States-9)]
Governor Word: [ROOT] Relation: [root] Dependent Word: [Who/WP]
Governor Word: [Who/WP] Relation: [copula] Dependent Word: [is/VBZ]
```

```
Governor Word: [president/NN] Relation: [determiner] Dependent Word:
[the/DT]
Governor Word: [president/NN] Relation: [adjectival modifier] Dependent
Word: [32nd/JJ]
Governor Word: [Who/WP] Relation: [nominal subject] Dependent Word:
[president/NN]
Governor Word: [States/NNPS] Relation: [determiner] Dependent Word: [the/
DT]
Governor Word: [States/NNPS] Relation: [nn modifier] Dependent Word:
[United/NNP]
Governor Word: [president/NN] Relation: [prep_collapsed] Dependent Word:
[States/NNPS]
```

这些信息是判断问题类型的基础。

7.6.2 判断问题类型

关系判断揭示出了问题类型判断的方法，比如去判断这是一个"who"的问题，我们应当验证关系是 nominal subject（名词性主语）且核心词是 who。

下面的代码中，我们遍历问题中的依赖关系类型及核心词是否相符，如果相符则调用 processWhoQuestion 方法进行处理：

```
for (TypedDependency dependency : tdl) {
    if ("nominal subject".equals( dependency.reln().getLongName())
        && "who".equalsIgnoreCase( dependency.gov().originalText())) {
        processWhoQuestion(tdl);
    }
}
```

这种简单的区分相当有效。它可以正确地识别同一问题下面的所有变化：

```
Who is the 32nd president of the United States?
Who was the 32nd president of the United States?
The 32nd president of the United States was who?
The 32nd president is who of the United States?
```

我们还可以使用不同的选择标准来判断其他问题类型。下面的问题会被分类为其他问题类型：

```
What was the 3rd President's party?
When was the 12th president inaugurated?
Where is the 30th president's home town?
```

我们可以通过下表中的关系判断问题类型：

问题类型	关系	核心词	依赖关系
what	名词性主语	what	无
when	状语	无	when
where	状语	无	where

这种方法确实需要硬编码关系。

7.6.3 搜索答案

一旦我们知道了问题的类型，就可以用文本中的关系来回答问题。为了说明这一过程，我们将使用 processWhoQuestion 方法。这种方法使用 TypedDependency 列表来存储所需的信息以回答关于总统的"who"类问题。具体来说，我们需要根据总统的任职顺序知道该问题对应的总统。

createPresidentList 方法生成一个列表用以查找相关信息。该方法读取文件 PresidentList 中总统的名字、就职及卸任年份，内容的格式如下：

George Washington   (1789-1797)

下面的 createPresidentList 方法使用 OpenNLP 的 SimpleTokenizer 类对每行进行分词，总统名字由不定个数的词项组成，确定总统名字后时间便容易提取：

```java
public List<President> createPresidentList() {
    ArrayList<President> list = new ArrayList<>();
    String line = null;
    try (FileReader reader = new FileReader("PresidentList");
            BufferedReader br = new BufferedReader(reader)) {
        while ((line = br.readLine()) != null) {
            SimpleTokenizer simpleTokenizer =
                SimpleTokenizer.INSTANCE;
            String tokens[] = simpleTokenizer.tokenize(line);
            String name = "";
            String start = "";
            String end = "";
            int i = 0;
            while (!"(".equals(tokens[i])) {
                name += tokens[i] + " ";
                i++;
            }
            start = tokens[i + 1];
            end = tokens[i + 3];
            if (end.equalsIgnoreCase("present")) {
```

```
                end = start;
            }
            list.add(new President(name,
                Integer.parseInt(start),
                Integer.parseInt(end)));
        }
    } catch (IOException ex) {
        // Handle exceptions
    }
    return list;
}
```

President 类存储了总统的信息,遗漏了 getter 方法,如下:

```
public class President {
    private String name;
    private int start;
    private int end;

    public President(String name, int start, int end) {
        this.name = name;
        this.start = start;
        this.end = end;
    }
    ...
}
```

下面是 processWhoQuestion 方法。我们继续使用类型依赖关系来提取问题的序数值。如果核心词是 president(总统)且其关系词是 adjectival modifier(形容词修饰语),那么依赖词就是序数词。将这一字符串传递给 getOrder 方法,可返回整数类型的序数词。结果要加上 1,因为总统的列表也是从 1 开始的:

```
public void processWhoQuestion(List<TypedDependency> tdl) {
    List<President> list = createPresidentList();
    for (TypedDependency dependency : tdl) {
        if ("president".equalsIgnoreCase(
                dependency.gov().originalText())
                && "adjectival modifier".equals(
                    dependency.reln().getLongName())) {
            String positionText =
                dependency.dep().originalText();
            int position = getOrder(positionText)-1;
            System.out.println("The president is "
                + list.get(position).getName());
        }
    }
}
```

getOrder 方法如下，仅仅是取出第一个数字字符并转换为整数。更复杂的方法可以是查找序数词的其他变种，如"第一"和"第十六"：

```
private static int getOrder(String position) {
    String tmp = "";
    int i = 0;
    while (Character.isDigit(position.charAt(i))) {
        tmp += position.charAt(i++);
    }
    return Integer.parseInt(tmp);
}
```

执行结果如下：

```
The president is Franklin D . Roosevelt
```

这一流程是提取问题信息并回答问题的一个简单的例子。其他类型的问题解决方法基本相同，留给读者作为练习。

7.7 本章小结

本章我们讨论了解析文本、提取信息的过程，可以应用于语法检查、机器翻译等一系列目的。文本内的关系有许多种，诸如父子关系、空间关系等，即文本内元素之间的相互关系。

文本解析可以得到文本内存在的关系，可以用来提取所需的信息。书中演示了许多应用 OpenNLP 和斯坦福大学的 API 进行文本解析的方法。

我们还说明了使用斯坦福大学的 API 判断文本内的共指消解，即两个或以上的表达式指向同一个人或物。

最后我们以问答系统为例，使用解析器从问题中提取关系，并用这些关系提取信息，用来回答关于美国总统的"who"类型的简单问题。

下一章，我们将介绍综合前七章所有的方法解决更复杂的问题。

第 8 章

方法组合

在这一章中，我们将使用技术的组合来解决自然语言处理中的几个问题。首先，我们简要介绍准备数据的过程。然后探讨流水线及流水线的构建。流水线无非就是组合一系列任务来解决一些问题。流水线的主要优点是能够插入和删除流水线中的各种元素，因此只需要很少的改动，就可以灵活地处理各种问题。

斯坦福大学的自然语言处理 API 能够很好地支持流水线结构，我们已经在本书中反复使用了这套工具。我们将会探索这种方法的更多细节，然后展示如何使用 OpenNLP 来构建一个流水线。

在自然语言处理问题中，准备待处理的数据是重要的第一步。我们在第 1 章中介绍了准备数据的过程，然后在第 2 章中讨论了归一化过程。在本章中，我们将重点放在从不同的数据源中提取文本，这些数据源主要包括 HTML、Word 和 PDF 文档。

斯坦福大学 API 中的 StanfordCoreNLP 类是流水线的一个很好的例子。从某种意义上说，它是预先构建的，实际执行的任务取决于对流水线添加的注解。这种使用流水线的方式适用于许多类型的问题。

然而，其他的 NLP API 并不像斯坦福大学 API 那样直接支持流水线结构；尽管构建更加困难，但是这些方法对于许多应用也显得更加灵活。我们将使用 OpenNLP 来展示流水线的构建过程。

8.1 准备数据

文本提取是 NLP 任务中的第一步。在这里，我们将快速介绍一下如何从 HTML、Word 和 PDF 文档中提取文本。虽然有多个 API 可以支持这些任务，但我们将主要使用以下的 API：

- 用于 HTML 文档的 Boilerpipe（https://code.google.com/p/boilerpipe/）
- 用于 Word 文档的 POI（http://poi.apache.org/index.html）
- 用于 PDF 文档的 PDFBox（http://pdfbox.apache.org/）

一些 API 还支持使用 XML 进行输入和输出。例如，斯坦福大学 API 中的 XMLUtils 类可以读取 XML 文件，操作 XML 数据。LingPipe 软件包中的 XMLParser 类也可以解析 XML 文本。

实际工作中，人们一般会采用多种形式来存储数据，并且这些数据通常不是简单的文本文件。演示文档存储在 PowerPoint 幻灯片中，说明书会使用 Word 文档创建，公司提供的市场营销材料常常是 PDF 文档。许多组织和机构还会在互联网上展示文档，这意味着许多有用的信息保存在 HTML 文件中。由于这些数据源的广泛性，我们需要使用工具来提取文本以便进行处理。

8.1.1 使用 Boilerpipe 从 HTML 中提取文本

我们可以使用一些库从 HTML 文档中提取文本。我们将演示如何使用 Boilerpipe 来提取文本。这是一个灵活的 API，它不仅可以提取 HTML 文档中的所有文本，也可以提取 HTML 文档中的特定部分，比如文档的标题或特定的文本块。

在演示 Boilerpipe 的时候，我们使用的 HTML 页面位于 http://en.wikipedia.org/wiki/Berlin。下图展示了这个网页的一部分。为了使用 Boilerpipe，你还需要下载 Xerces Parser 的二进制代码，这些代码可以从 http://xerces.apache.org/index.html 上找到。

我们首先创建一个表示这个页面的 URL 对象，代码如下所示，其中的 try-catch 代码块用来处理异常：

```
try {
    URL url = new
```

```
URL("http://en.wikipedia.org/wiki/Berlin");
…
} catch (MalformedURLException ex) {
    // Handle exceptions
} catch (BoilerpipeProcessingException | SAXException
        | IOException ex) {
    // Handle exceptions
}
```

我们将使用两个类来提取文本。第一个是 HTMLDocument 类,它用来表示 HTML 文档。第二个是 TextDocument 类,它表示 HTML 文档中的文本。TextDocument 中可以包含一个或多个 TextBlock 对象,这些对象可以根据需要进行单独访问。

下一步,我们给维基百科柏林网页创建一个 HTMLDocument 实例。Boilerpipe-SAXInput 类使用该输入源创建一个 TextDocument 实例。随后使用 TextDocument 类中的 getText 方法获取文本。getText 方法有两个参数。第一个参数指定了是否包含标注为内容的 TextBlock 实例。第二个参数指定了是否包含非内容的 TextBlock 实例。在这个例子中，这两种类型的 TextBlock 实例都需要被包含：

```
HTMLDocument htmlDoc = HTMLFetcher.fetch(url);
InputSource is = htmlDoc.toInputSource();
TextDocument document =
    new BoilerpipeSAXInput(is).getTextDocument();
System.out.println(document.getText(true, true));
```

因为原网页很大，所以这一段代码的输出是很庞大的。程序输出的一部分如下所示：

```
Berlin
From Wikipedia, the free encyclopedia
Jump to: navigation , search
This article is about the capital of Germany.  For other uses, see Berlin
(disambiguation) .
...
Privacy policy
About Wikipedia
Disclaimers
Contact Wikipedia
Developers
Mobile view
```

getTextBlocks 方法将会返回一个包含 TextBlock 对象的列表。除了可以获取文本之外，也有一些方法可以获取文本的信息，比如一块文本中的字符个数。

8.1.2　使用 POI 从 Word 文档中提取文本

Apache POI 项目（http://poi.apache.org/index.html）提供的 API 可以用来从微软 Office 产品中提取信息。这个扩展库能够从 Word 文档和其他 Office 产品中提取信息，如 Excel 和 Outlook。

当下载使用 POI 库时，还需要用到 XMLBeans（http://xmlbeans.apache.org/）。XMLBeans 的二进制文件可以从站点 http://www.java2s.com/Code/Jar/x/Downloadxml-beans230jar.htm 获得。

我们的主要关注点在于如何使用 POI 从 Word 文档中提取文本。为了演示 POI 的使用，我们将使用一个名为 TestDocument.docx 的文档，下图显示了该文档的内容：

> **Pirates**
>
> Pirates are people who use ships to rob other ships. At least this is a common definition. They have also been known as buccaneers, corsairs, and privateers. In recent times, the term has been expanded to include all sorts of villains including people who pirate software.
>
> **List of Historical Pirates**
>
> This is not intended to be a complete list. A fuller list can be found at http://en.wikipedia.org/wiki/List_of_pirates. Our list includes:
>
> - Gan Ning
> - Awilda
> - John Crabbe
> - Sir Francis Drake (As least to the Spanish)
> - Blackbeard (Edward Teach)
> - "Calico Jack" John Rackham
> - Chui A-poo
> - Johnny Depp (Opps, acted as a pirate)
>
> **How to become a Pirate**
>
> For those of you who have the inclination, the following is one approach to become a pirate:
>
> 1. Recruit fellow scurvy dogs
> 2. Steal a ship
> 3. Plunder the high seas
> 4. Get caught
> 5. Walk the plank
>
> This is not a recommended occupation.

不同的 Word 版本使用的文件格式不一样。为了简化文本提取类的选择，我们使用 ExtractorFactory 工厂类。

尽管 POI 的功能非常强大，提取文本的过程却很简单。正如下面程序展示的那样，FileInputStream 对象表示文档 TestDocument.docx，然后使用 ExtractorFactory 类中的 createExtractor 方法来选择适当的 POITextExtractor 实例。POITextExtractor 类是几个不同提取器的基类。最后再用 getText 方法从提取器中获取文本：

```java
try {
    FileInputStream fis =
        new FileInputStream("TestDocument.docx");
    POITextExtractor textExtractor =
        ExtractorFactory.createExtractor(fis);
    System.out.println(textExtractor.getText());
} catch (IOException ex) {
    // Handle exceptions
```

```
} catch (OpenXML4JException | XmlException ex) {
    // Handle exceptions
}
```

程序的部分输出如下所示:

```
Pirates
Pirates are people who use ships to rob other ships. At least this is a
common definition. They have also been known as buccaneers, corsairs, and
privateers. In
...
Our list includes:
Gan Ning
Awilda
...
Get caught
Walk the plank
This is not a recommended occupation.
```

当需要了解更多的 Word 文档属性时,POI 中的 POIXMLPropertiesTextExtractor 类提供了访问文档的核心属性、扩展属性和自定义属性的方法。有两种方法可以轻易地得到一个包含许多属性的字符串。

- 第一种方法是先使用 getMetadataTextExtractor 方法,然后再使用 getText 方法。代码如下:

```
POITextExtractor metaExtractor =
    textExtractor.getMetadataTextExtractor();
System.out.println(metaExtractor.getText());
```

- 第二种方法是先使用 XWPFDocument 表示这个 Word 文档,然后再创建一个 POIXMLPropertiesTextExtractor 类的实例。代码如下:

```
fis = new FileInputStream("TestDocument.docx");
POIXMLPropertiesTextExtractor properties =
    new POIXMLPropertiesTextExtractor(new
    XWPFDocument(fis));
System.out.println(properties.getText());
```

无论选择哪种方法,它们的结果都如下所示:

```
Created = Sat Jan 03 18:27:00 CST 2015
CreatedString = 2015-01-04T00:27:00Z
Creator = Richard
LastModifiedBy = Richard
```

```
LastPrinted = Sat Jan 03 18:27:00 CST 2015
LastPrintedString = 2015-01-04T00:27:00Z
Modified = Mon Jan 05 14:01:00 CST 2015
ModifiedString = 2015-01-05T20:01:00Z
Revision = 3
Application = Microsoft Office Word
AppVersion = 12.0000
Characters = 762
CharactersWithSpaces = 894
Company =
HyperlinksChanged = false
Lines = 6
LinksUpToDate = false
Pages = 1
Paragraphs = 1
Template = Normal.dotm
TotalTime = 20
```

CoreProperties 类可以存放文档的核心属性。getCoreProperties 方法可以访问这些属性:

```
CoreProperties coreProperties = properties.getCoreProperties();
System.out.println(properties.getCorePropertiesText());
```

文档的核心属性如下所示:

```
Created = Sat Jan 03 18:27:00 CST 2015
CreatedString = 2015-01-04T00:27:00Z
Creator = Richard
LastModifiedBy = Richard
LastPrinted = Sat Jan 03 18:27:00 CST 2015
LastPrintedString = 2015-01-04T00:27:00Z
Modified = Mon Jan 05 14:01:00 CST 2015
ModifiedString = 2015-01-05T20:01:00Z
Revision = 3
```

如果需要访问特定的属性,我们可以使用一些单独的方法,比如 getCreator、getCreated 和 getModified。扩展属性可以由 ExtendedProperties 类表示,可以使用 getExtendedProperties 方法访问文档的扩展属性,代码如下:

```
ExtendedProperties extendedProperties =
    properties.getExtendedProperties();
System.out.println(properties.getExtendedPropertiesText());
```

文档的扩展属性输出为:

```
Application = Microsoft Office Word
AppVersion = 12.0000
Characters = 762
CharactersWithSpaces = 894
Company =
HyperlinksChanged = false
Lines = 6
LinksUpToDate = false
Pages = 1
Paragraphs = 1
Template = Normal.dotm
TotalTime = 20
```

如果需要访问文档特定的扩展属性,可以使用 getApplication、getAppVersion 和 getPages 等方法。

8.1.3 使用 PDFBox 从 PDF 文档中提取文本

Apache PDFBox 项目提供了处理 PDF 文档的 API。PDFBox 除了可以提取文本,还可以完成其他任务,这些任务包括合并文档,填写表格,创建 PDF 等。在这里,我们仅仅演示文本提取过程。

为了演示 PDFBox 的使用,我们将用到一个名为 TestDocument.pdf 的文档。这个 PDF 文档是由上一节中的 TestDocument.docx 文件另存为得到的。

处理过程很简单。首先为 PDF 文档创建一个 File 对象,然后使用 PDDocument 类表示这个文档,最后使用 PDFTextStripper 类中的 getText 方法进行实际的文本提取工作。代码如下所示:

```java
try {
    File file = new File("TestDocument.pdf");
    PDDocument pdDocument = PDDocument.load(file);
    PDFTextStripper stripper = new PDFTextStripper();
    String text = stripper.getText(pdDocument);
    System.out.println(text);
    pdDocument.close();
} catch (IOException ex) {
    // Handle exceptions
}
```

由于原文档较长,这里只给出部分输出结果:

```
Pirates
Pirates are people who use ships to rob other ships. At least this is a
common definition. They have also been known as buccaneers, corsairs, and
privateers. In
...
Our list includes:
☒ Gan Ning
☒ Awilda
...
4. Get caught
5. Walk the plank
This is not a recommended occupation.
```

这种方法还可以提取出文档中编号列表项前面的数字和特殊的编号标记字符。

8.2 流水线

流水线可以看作一系列的操作，其中每一个操作的输出都被用作另一个操作的输入。我们在前面章节里已经看到了流水线的几个例子，但是这些例子都比较短。尤其是我们看到了斯坦福大学 API 中的 StanfordCoreNLP 类，结合注解的使用，很好地支持了流水线这一概念。我们将在下一节讨论这种方法。

流水线的优点之一是，如果构建得当，它可以方便地添加和删除处理元件。例如，如果流水线中的某一个步骤是将字符转换为小写，那么我们也可以轻易地删除这一步骤，与此同时，流水线的其余元件保持不变。

然而，也有些流水线没有这么灵活，某一步可能需要上一步才能正常工作。在由 StanfordCoreNLP 类支持的流水线中，为了支持 POS 处理，还需要加上一些注解：

```
props.put("annotators", "tokenize, ssplit, pos");
```

如果我们删除 ssplit 注解，程序将会出现一个异常：

```
java.lang.IllegalArgumentException: annotator "pos" requires
annotator "ssplit"
```

在斯坦福大学 API 中，我们不需要花费很多精力来构建流水线，但是其他的 API 可能需要花费大量精力。我们将在 8.3 节中介绍复杂流水线的构建。

8.2.1 使用 Stanford 流水线

在本节中，我们将讨论 Stanford 流水线的更多细节。虽然我们在这本书的几个例子中使用过流水线，但是还没有充分挖掘出它的功能。通过之前的使用，你现在可以更好地理解如何使用流水线。阅读完这部分以后，你将会更深入地了解到流水线的功能和适用范围。

edu.stanford.nlp.pipeline 程序包中包含了 StanfordCoreNLP 类和 annotator 类。流水线通常的使用方法如下所示。这段代码处理的是一个名为 text 的字符串。Properties 类中包含了注释的名称：

```
String text = "The robber took the cash and ran.";
Properties props = new Properties();
props.put("annotators",
    "tokenize, ssplit, pos, lemma, ner, parse, dcoref");
StanfordCoreNLP pipeline = new StanfordCoreNLP(props);
```

Annotation 类表示待处理的文本。它的构造器使用待处理的字符串作为参数，为 Annotation 对象添加一个 CoreAnnotations.TextAnnotation 实例。StanfordCoreNLP 类中的 annotate 方法可以将前面属性列表中指定的注释应用到 Annotation 对象上：

```
Annotation annotation = new Annotation(text);
pipeline.annotate(annotation);
```

CoreMap 接口是所有注释对象的基接口。它使用类对象作为键。TextAnnotation 注释类型就是一个表示文本的 CoreMap 键。CoreMap 键可以与各种类型的注释一起使用，比如那些在属性列表中定义的注释。值取决于键的类型。

下图描述了类和接口的层次结构。它是类和接口之间关系的简化版本。水平线代表接口的实现，而垂直线表示类之间的继承。

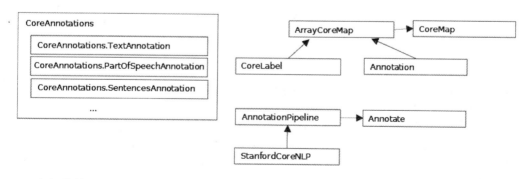

我们将使用下面的代码来验证 annotate 方法的效果。keyset 方法将返回一个集合，该集合中包含了 Annotation 对象中所有的注释键。程序将分别显示 annotate 方法应用前和应用后键的值。

```
System.out.println("Before annotate method executed ");
Set<Class<?>> annotationSet = annotation.keySet();
for(Class c : annotationSet) {
    System.out.println("\tClass: " + c.getName());
}

pipeline.annotate(annotation);

System.out.println("After annotate method executed ");
annotationSet = annotation.keySet();
for(Class c : annotationSet) {
    System.out.println("\tClass: " + c.getName());
}
```

程序的输出结果如下。结果显示：Annotation 对象的创建使得 TextAnnotation 扩展被添加到注释中。执行 annotate 方法以后，还有一些附加的注释也被添加了进去：

```
Before annotate method executed
    Class: edu.stanford.nlp.ling.CoreAnnotations.TextAnnotation
After annotate method executed
    Class: edu.stanford.nlp.ling.CoreAnnotations.TextAnnotation
    Class: edu.stanford.nlp.ling.CoreAnnotations.TokensAnnotation
    Class: edu.stanford.nlp.ling.CoreAnnotations.SentencesAnnotation
    Class: edu.stanford.nlp.dcoref.CorefCoreAnnotations.
CorefChainAnnotation
```

CoreLabel 类实现了 CoreMap 接口。它代表一个单独的词，注释信息附加在这个词上。所附的信息取决于流水线创建时的属性集。然而，还可以使用一些位置信息，比如它的开始和结束位置或者实体前后的空格。

CoreMap 或者 CoreLabel 中的 get 方法的返回信息由它的参数确定。重载的 get 方法返回值取决于参数类型。举个例子，下面是 SentencesAnnotation 类的声明，它实现了 CoreAnnotation<List<CoreMap>>：

```
public static class CoreAnnotations.SentencesAnnotation
    extends Object
    implements CoreAnnotation<List<CoreMap>>
```

下列语句中，SentencesAnnotation 类返回的是一个 List<CoreMap> 实例：

```
List<CoreMap> sentences =
    annotation.get(SentencesAnnotation.class);
```

类似地，TokensAnnotation 类实现了 CoreAnnotation<List<CoreLabel>>：

```
public static class CoreAnnotations.TokensAnnotation
    extends Object
    implements CoreAnnotation<List<CoreLabel>>
```

它的 get 方法返回的是一个 List<CoreLabel> 实例,并在 for-each 语句中使用:

```
for (CoreLabel token : sentence.get(TokensAnnotation.class)) {
```

在前面的章节中,我们已经使用了 SentencesAnnotation 类来访问一个注释中的句子:

```
List<CoreMap> sentences =
    annotation.get(SentencesAnnotation.class);
```

CoreLabel 类用来访问一个句子中单个的词,具体代码如下:

```
for (CoreMap sentence : sentences) {
    for (CoreLabel token:
            sentence.get(TokensAnnotation.class)) {
        String word = token.get(TextAnnotation.class);
        String pos = token.get(PartOfSpeechAnnotation.class);
    }
}
```

Annotator 的参数选项可以查阅 http://nlp.stanford.edu/software/corenlp.shtml。下面的代码演示了如何使用一个注释来指定 POS 模型。使用 Property 类中的 put 方法来设置模型中的 pos.model 属性:

```
props.put("pos.model",
    "C:/.../Models/english-caseless-left3words-distsim.tagger");
```

下表总结了注释的使用方法:第一列是属性列表中的字符串,第二列只列出了基本的注解类,第三列是它们典型的使用方法。

属性名称	基本的注释类	使用方法
tokenize	TokensAnnotation	分词
cleanxml	XmlContextAnnotation	移除 XML 标记
ssplit	SentencesAnnotation	把词项拆分成句子
pos	PartOfSpeechAnnotation	创建 POS 标注
lemma	LemmaAnnotation	词形还原
ner	NamedEntityTagAnnotation	创建 NER 标注
regexner	NamedEntityTagAnnotation	使用正则表达式创建 NER 标注
sentiment	SentimentCoreAnnotations	情感分析
truecase	TrueCaseAnnotation	真实案例分析
parse	TreeAnnotation	生成解析树
Depparse	BasicDependenciesAnnotation	句法依存分析
dcoref	CorefChainAnnotation	共指消解
relation	MachineReadingAnnotations	关系提取

使用下面的代码创建一个流水线：

```
String text = "The robber took the cash and ran.";
Properties props = new Properties();
props.put("annotators",
    "tokenize, ssplit, pos, lemma, ner, parse, dcoref");
StanfordCoreNLP pipeline = new StanfordCoreNLP(props);
```

以下结果展示了注释添加的过程。我们可以看到每个注释添加时的情况：

```
Adding annotator tokenize
TokenizerAnnotator: No tokenizer type provided. Defaulting to
PTBTokenizer.
Adding annotator ssplit
edu.stanford.nlp.pipeline.AnnotatorImplementations:
Adding annotator pos
Reading POS tagger model from edu/stanford/nlp/models/pos-tagger/english-
left3words/english-left3words-distsim.tagger ... done [2.5 sec].
Adding annotator lemma
Adding annotator ner
Loading classifier from edu/stanford/nlp/models/ner/english.all.3class.
distsim.crf.ser.gz ... done [6.7 sec].
Loading classifier from edu/stanford/nlp/models/ner/english.muc.7class.
distsim.crf.ser.gz ... done [5.0 sec].
Loading classifier from edu/stanford/nlp/models/ner/english.conll.4class.
distsim.crf.ser.gz ... done [5.5 sec].
Adding annotator parse
Loading parser from serialized file edu/stanford/nlp/models/lexparser/
englishPCFG.ser.gz ...done [0.9 sec].
Adding annotator dcoref
```

当使用 annotate 方法时，我们可以使用 timingInformation 方法来查看各个步骤所花费的时间，代码如下：

```
System.out.println("Total time: " + pipeline.timingInformation());
```

结果如下：

```
Total time: Annotation pipeline timing information:
TokenizerAnnotator: 0.0 sec.
WordsToSentencesAnnotator: 0.0 sec.
POSTaggerAnnotator: 0.0 sec.
MorphaAnnotator: 0.1 sec.
NERCombinerAnnotator: 0.0 sec.
ParserAnnotator: 2.5 sec.
```

```
DeterministicCorefAnnotator: 0.1 sec.
TOTAL: 2.8 sec. for 8 tokens at 2.9 tokens/sec.
```

8.2.2 在 Standford 流水线中使用多核处理器

annotate 方法能够充分利用多核处理器。它是一个重载的方法，其中有一个重载的版本可以用 Iterable<Annotation> 实例作为参数。这个方法将充分利用可用的处理器来处理 Annotation 实例。

我们使用前面已经定义的 pipeline 对象来展示这个版本的 annotate 方法。

首先，我们用 4 个短句子创建 4 个 Annotation 对象。为了充分利用这项技术，最好使用一个更大的数据集：

```
Annotation annotation1 = new Annotation(
    "The robber took the cash and ran.");
Annotation annotation2 = new Annotation(
    "The policeman chased him down the street.");
Annotation annotation3 = new Annotation(
    "A passerby, watching the action, tripped the thief "
    + "as he passed by.");
Annotation annotation4 = new Annotation(
    "They all lived happily ever after, except for the thief "
    + "of course.");
```

ArrayList 类实现了 Iterable 接口。我们先创建一个 ArrayList 实例，然后把 4 个 Annotation 对象添加进去，将列表分配给一个 Iterable 变量：

```
ArrayList<Annotation> list = new ArrayList();
list.add(annotation1);
list.add(annotation2);
list.add(annotation3);
list.add(annotation4);
Iterable<Annotation> iterable = list;
```

随后执行 annotate 方法：

```
pipeline.annotate(iterable);
```

我们将使用 annotation2 方法来展示每个词及其 POS：

```
List<CoreMap> sentences =
    annotation2.get(SentencesAnnotation.class);

for (CoreMap sentence : sentences) {
    for (CoreLabel token :
            sentence.get(TokensAnnotation.class)) {
```

```
            String word = token.get(TextAnnotation.class);
            String pos = token.get(PartOfSpeechAnnotation.class);
            System.out.println("Word: " + word + " POS Tag: " + pos);
        }
    }
```

结果如下：

```
Word: The POS Tag: DT
Word: policeman POS Tag: NN
Word: chased POS Tag: VBD
Word: him POS Tag: PRP
Word: down POS Tag: RP
Word: the POS Tag: DT
Word: street POS Tag: NN
Word: . POS Tag: .
```

上面的代码表明，Standford 流水线可以轻易地实现并行处理。

8.3 创建一个文本搜索的流水线

搜索是一个丰富且复杂的话题。为了执行一次搜索操作，我们可能需要用到各种类型的搜索算法。在这里，我们将展示如何使用各种各样的 NLP 技术来实现文本搜索。

在大多数机器上，单个的文本文档可以在合理的时间内得到处理。然而，当需要搜索多个大型文档时，常见的方法是创建一个索引。这种方法可以让搜索过程在合理的时间内完成。

我们将演示创建索引的一种方法，然后使用索引进行搜索操作。我们这里使用的文本不是很大，但也足以演示这一过程了。

我们需要做的是：

1）从文件中读取文本

2）分词并判断句子的边界

3）去除停用词

4）累计索引统计值

5）写入索引文件

下面几个因素会影响到索引文件的内容：

- 去除停用词
- 区分大小写的搜索
- 寻找同义词
- 词干和词形还原
- 允许跨越句子边界搜索

我们将使用 OpenNLP 来演示这个过程。这个例子的目的是为了说明如何在一个流水线中组合多种 NLP 技术来解决搜索类型的问题。这不是一个全面的解决方案，我们将忽略一些技术，比如词干提取。此外，这里不会实际地创建一个索引文件，而是将它作为练习留给读者。在这里，我们将专注于如何使用 NLP 技术。

具体而言，我们将：

- 把书拆分成句子
- 把句子转换为小写
- 去除停用词
- 创建内部索引数据结构

我们将开发两个类来支持索引数据结构：Word 类和 Positions 类。我们将增强 StopWords 类，这个类在第 2 章中已经开发完成，在这里我们让它支持 removeStopWords 方法的一个重载版本。这个新版本的方法将能更方便地去除停用词。

我们首先使用 try-with-resources 语句块来打开一个文件输入流，用文件输入流来读取句子模型，句子模型保存在 en-sent.bin 中。然后再打开待处理的文件，文件内容是儒勒·凡尔纳的《海底两万里》。这本书是从 http://www.gutenberg.org/ebooks/164 下载的，并且做了一些修改，删除了开头和结尾的谷登堡文本（Gutenberg text），使其更易于阅读：

```
try (InputStream is = new FileInputStream(new File(
    "C:/Current Books/NLP and Java/Models/en-sent.bin"));
    FileReader fr = new FileReader("Twenty Thousands.txt");
    BufferedReader br = new BufferedReader(fr)) {
        …
} catch (IOException ex) {
    // Handle exceptions
}
```

用句子模型创建一个 SentenceDetectorME 类的实例，如下所示：

```
SentenceModel model = new SentenceModel(is);
SentenceDetectorME detector = new SentenceDetectorME(model);
```

接下来，我们使用 StringBuilder 实例来创建一个字符串，这个字符串用来支持句子边界检测。读取这本书并将它添加到 StringBuilder 实例。然后使用 sentDetect 方法创建句子数组，如下所示：

```
String line;
StringBuilder sb = new StringBuilder();
while ((line = br.readLine()) != null) {
    sb.append(line + " ");
}
String sentences[] = detector.sentDetect(sb.toString());
```

对于修改版本的《海底两万里》文件，这种方法创建了一个包含 14 859 个句子的数组。

接下来，我们使用 toLowerCase 方法将文本全部转换为小写。这样做的原因是：在去除停用词时，这样的处理可以捕获所有的停用词。

```
for (int i = 0; i < sentences.length; i++) {
    sentences[i] = sentences[i].toLowerCase();
}
```

转换为小写和去除停用词会对搜索产生一些限制。然而，这可以认为是这种实现方法的一个特点，在其他实现方法中可以做出一定的调整。

接下来，去除停用词。正如前面所说的那样，我们将添加 removeStopWords 方法的一个重载版本，这个重载版本将简化操作。新方法代码如下：

```
public String removeStopWords(String words) {
    String arr[] =
        WhitespaceTokenizer.INSTANCE.tokenize(words);
    StringBuilder sb = new StringBuilder();
    for (int i = 0; i < arr.length; i++) {
        if (stopWords.contains(arr[i])) {
            // Do nothing
        } else {
            sb.append(arr[i]+" ");
        }
    }
    return sb.toString();
}
```

我们使用 stop-words_english_2_en.txt 文件创建一个 StopWords 实例。这个文件可以从 https://code.google.com/p/stop-words/ 下载。我们之所以选择这个文件，是因为我们认

为这个文件中的停用词适合于这本书。

```
StopWords stopWords = new StopWords("stop-words_english_2_en.txt");
for (int i = 0; i < sentences.length; i++) {
    sentences[i] = stopWords.removeStopWords(sentences[i]);
}
```

到了这一步，文本已经处理完毕。下一步，我们将在已处理文本的基础上，创建一个索引数据结构。索引数据结构将使用 Word 类和 Positions 类。Word 类中定义了两个成员变量：一个是代表单词的成员变量 Word，另一个成员变量是内容为 Positions 对象的动态数组（ArrayList）。一个文档中，相同的单词可能会出现多次，因此使用动态数组来保存单词在文档中出现的位置。这个类定义如下：

```
public class Word {
    private String word;
    private final ArrayList<Positions> positions;

    public Word() {
        this.positions = new ArrayList();
    }

    public void addWord(String word, int sentence,
            int position) {
        this.word = word;
        Positions counts = new Positions(sentence, position);
        positions.add(counts);
    }

    public ArrayList<Positions> getPositions() {
        return positions;
    }

    public String getWord() {
        return word;
    }
}
```

Positions 类中定义了两个成员变量：sentence 代表句子的数量，position 代表单词在句子中出现的位置。这个类定义如下：

```
class Positions {
    int sentence;
    int position;

    Positions(int sentence, int position) {
        this.sentence = sentence;
        this.position = position;
    }
}
```

为了使用这两个类，我们需要创建一个 HashMap 实例来保存文件中每个单词的位置信息：

```
HashMap<String, Word> wordMap = new HashMap();
```

HashMap 中 Word 的创建方法如下所示。我们先将句子进行分词，然后分别检查词项是否在 HashMap 中。HashMap 中把单词用作键。

我们用 containsKey 方法来判断单词是否已经添加到 HashMap 中。如果已经添加了，那么删除之前的 Word 实例。如果没有添加，那么就创建一个新的 Word 实例。不论怎样，新的位置信息被附加到 Word 实例中，然后将 Word 实例添加到 HashMap 中：

```
for (int sentenceIndex = 0;
        sentenceIndex < sentences.length; sentenceIndex++) {
    String words[] = WhitespaceTokenizer.INSTANCE.tokenize(
        sentences[sentenceIndex]);
    Word word;
    for (int wordIndex = 0;
            wordIndex < words.length; wordIndex++) {
        String newWord = words[wordIndex];
        if (wordMap.containsKey(newWord)) {
            word = wordMap.remove(newWord);
        } else {
            word = new Word();
        }
        word.addWord(newWord, sentenceIndex, wordIndex);
        wordMap.put(newWord, word);
    }
}
```

为了演示实际的查找过程，我们使用 get 方法来返回单词"reef"对应的 Word 对象。然后使用 getPositions 方法来获得位置列表，最终显示每个位置，代码如下：

```
Word word = wordMap.get("reef");
ArrayList<Positions> positions = word.getPositions();
for (Positions position : positions) {
    System.out.println(word.getWord() + " is found at line "
        + position.sentence + ", word "
        + position.position);
}
```

结果如下：

```
reef is found at line 0, word 10
reef is found at line 29, word 6
reef is found at line 1885, word 8
reef is found at line 2062, word 12
```

这个实现比较简单，但也演示了如何组合不同的 NLP 技术来创建和使用索引数据结构，最后获得的索引数据结构可以保存为一个索引文件。我们可以使用以下方法来增强搜索的性能：

- 其他的过滤操作
- 在 Positions 类中保存文档信息
- 在 Positions 类中保存章节信息
- 提供搜索选项，比如：
  - 区分大小写的搜索
  - 精确文本搜索
- 更好的异常处理

我们把这些留给读者当作练习。

8.4 本章小结

在本章中，我们讨论了数据的准备过程和流水线的创建过程，我们还演示了从 HTML、Word 和 PDF 文档中提取文本的方法。

我们可以看到，流水线无非就是组合一系列的任务来解决问题。我们可以根据需要插入和删除流水线中的各种元素。我们详细探讨了 Standford 流水线的结构。此外，还验证了不同注释的使用方法。在学习使用多核处理器的时候，我们探索了流水线的更多细节。

我们演示了如何使用 OpenNLP 构建一个流水线，并创建和使用索引进行文本搜索。这是另一种构建流水线的方法，与 Standford 流水线相比，这种构建方法允许更多的变化。

本书中，我们介绍了使用 Java 进行 NLP。我们覆盖了 NLP 中所有重要的任务，并且演示了如何使用各种各样的 NLP 的 API 来实现这些任务。NLP 这个领域纷繁复杂、莫可名状，我们祝愿你在今后的程序开发过程中一帆风顺。

推荐阅读

Java核心技术 卷I：基础知识（原书第10版）

书号：978-7-111-54742-6　作者：（美）凯 S. 霍斯特曼（Cay S. Horstmann）　定价：119.00元

  Java领域最有影响力和价值的著作之一，与《Java编程思想》齐名，10余年全球畅销不衰，广受好评

  根据Java SE 8全面更新，系统全面讲解Java语言的核心概念、语法、重要特性和开发方法，包含大量案例，实践性强

  本书为专业程序员解决实际问题而写，可以帮助你深入了解Java语言和库。在卷I中，Horstmann主要强调基本语言概念和现代用户界面编程基础，深入介绍了从Java面向对象编程到泛型、集合、lambda表达式、Swing UI设计以及并发和函数式编程的最新方法等内容。

推荐阅读

Python机器学习
作者：Sebastian Raschka, Vahid Mirjalili ISBN：978-7-111-55880-4 定价：79.00元

机器学习：实用案例解析
作者：Drew Conway, John Myles White ISBN：978-7-111-41731-6 定价：69.00元

面向机器学习的自然语言标注
作者：James Pustejovsky, Amber Stubbs ISBN：978-7-111-55515-5 定价：79.00元

机器学习系统设计：Python语言实现
作者：David Julian ISBN：978-7-111-56945-9 定价：59.00元

Scala机器学习
作者：Alexander Kozlov ISBN：978-7-111-57215-2 定价：59.00元

R语言机器学习：实用案例分析
作者：Dipanjan Sarkar, Raghav Bali ISBN：978-7-111-56590-1 定价：59.00元

推荐阅读

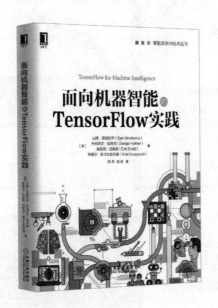 

面向机器智能的TensorFlow实践

作者:Sam Abrahams, Danijar Hafner, Erik Erwitt, Dan Becker ISBN:978-7-111-56389-1 定价:69.00元

本书是一本绝佳的TensorFlow入门指南。几位作者都来自谷歌研发一线,他们用自己的宝贵经验,结合众多高质量的代码,生动讲解TensorFlow的底层原理,并从实践角度介绍如何将两种常见模型——深度卷积网络、循环神经网络应用到图像理解和自然语言处理的典型任务中。此外,还介绍了在模型部署和编程中可用的诸多实用技巧。

TensorFlow机器学习实战指南

作者:Nick McClure ISBN:978-7-111-57948-9 定价:69.00元

本书由资深数据科学家撰写,从实战角度系统讲解TensorFlow基本概念及各种应用实践。真实的应用场景和数据,丰富的代码实例,详尽的操作步骤,带你由浅入深系统掌握TensorFlow机器学习算法及其实现。